Management for Professionals

For further volumes:
http://www.springer.com/series/10101

Christoph Karrer

Engineering Production Control Strategies

A Guide to Tailor Strategies that Unite
the Merits of Push and Pull

 Springer

Christoph Karrer
McKinsey & Company, Inc.
Sophienstr. 26
80333 München
Germany

Email: Christoph.Karrer@gmail.com

ISSN 2192-8096   e-ISSN 2192-810X
ISBN 978-3-642-24141-3   e-ISBN 978-3-642-24142-0
DOI 10.1007/978-3-642-24142-0
Springer Heidelberg Dordrecht London New York

Library of Congress Control Number: 2012933113

Printed on acid-free paper

Springer is part of Springer Science+Business Media (www.springer.com)

Foreword

Today's global supply chains are getting more and more complex. At the same time, the demand on customer service and the cost pressure continue to increase. In this challenging environment, production control strategies (PCS) play a major role. They manage the physical material flow on the shopfloor and are therefore a key driver for delivery performance, inventory levels, and ultimately production cost.

Identifying and customizing suitable control strategies is a challenging task, especially when production systems have to cope with variable demands, forecast error, and unstable processes.

The focus of this book lies on helping companies with complex and discrete production systems to tailor a production control strategy to their needs. Thereby, the mutual merits of "push" and "pull" systems are taken into account, leading to hybrid strategies. Consequently, the book addresses practitioners who are interested in looking behind the scenes and into the physics of production control.

A real-life case study demonstrates the practical applicability of the presented framework. I would like to thank the company and the involved managers for enabling this cooperation.

Moreover, I would like to express my gratitude to Prof. Dr. Hans-Otto Günther and PD Dr.-Ing. Knut Alicke for their support, our fruitful discussions, and for being inspirational mentors.

I would also like to thank my family for their constant support and understanding during the creation of this book. I dedicate this book to them.

Munich, Germany   Christoph Karrer

Customers require on-time delivery at a minimal cost. As a result, companies are constantly under pressure to cut costs and uphold high levels of service. A major factor for success in achieving these objectives is the right production control strategy, one in which two approaches compete in practical application – push and pull.

In push systems, external signals trigger production orders; these signals typically take the form of sophisticated, detailed, planning and scheduling algorithms in environments with integrated planning systems. A pull system initiates a production order internally. The consumption of parts of the next step in the value chain triggers the release of a signal (for example, a Kanban card), which is then translated into a production order.

Both production control approaches have clear benefits and disadvantages, and should, therefore, be combined. Many companies have already implemented pull systems as a part of their Lean manufacturing philosophy; this control strategy is easy to apply, and it limits work in process. Companies with strong planning systems prefer push systems to, for example, leverage the forecast. Unfortunately the push and pull approaches are often applied in a very dogmatic way that does not capture the benefits of either.

In his excellent book, Christoph Karrer presents a method for combining the two approaches. He provides a sound theoretical foundation for verifying the benefits of using a fraction of the existing forecast to control the production system. The results are promising; cost and inventory can be reduced significantly, and a high level of service retained. The beauty of his approach is its relative simplicity in practice – there is no need for system investment or radical changes in production control. In addition, the Kanban system – often already in place – can be leveraged in order to implement the approach.

Karrer's book is aimed at practitioners who contend with high fluctuations in demand and who would like to further reduce their costs after implementing a lean or an integrated planning system. His approach is a breakthrough – it combines Lean manufacturing ("pull") and "algorithmic" detailed planning and scheduling ("push"), and will further boost system performance.

Karlsruhe, Germany   PD Dr.-Ing. Knut Alicke
                                    Master Expert Supply Chain Management
                                            McKinsey & Company, Stuttgart

The quest for a good production control strategy (PCS) is as old as industrial production. Extensive research in the field has led to many innovations that enable today's production systems. The availability of affordable computer technology, which led to the introduction of IT-based planning systems, was an important milestone. Another important step was marked by the diffusion of the Lean manufacturing philosophy from Toyota, comprising the famous Kanban control system. However, due to the large variety of existing control strategies and the complexity of today's industrial practice, it is difficult for practioners to select and continuously update their PCS.

The engineering framework presented in this book offers valuable support. The strength of the approach is its integrated and practice oriented perspective. The problem is approached from a systems engineering angle, taking findings from current research into account. The resulting strategies combine merits of "push" and "pull" systems and yet remain in line with the philosophy of Lean manufacturing.

Berlin, Germany   Prof. Dr. Hans-Otto Günther
                                    Department of Production Management
                                            Technical University of Berlin

Contents

Chapter 1
Introduction

1.1 Need for a PCS Engineering Framework

Today's manufacturing reality is characterized by a constantly increasing complexity of production systems. The growing complexity of products, shorter product lifecycles, mass customization,[1] or agile manufacturing[2] are some of the driving factors behind this change (Abele et al. 2009; Hopp and Spearman 2008). This development imposes high demands on the production control strategy. To achieve high delivery performance, quality, and flexibility at lowest cost possible in complex production systems, it is essential to have the right production control strategy (PCS) in place.

As a rule of thumb, it can be stated that the more complex a production system becomes, the larger is the impact of the PCS on operational performance and thus the competiveness of the company. However, in many instances, shopfloor reality shows that a proper solution is still overdue. Especially in complex discrete manufacturing like electronics production, often neither the logic provided by the standard Enterprise Resource Planning (ERP) systems, nor the production control approaches offered by Lean Manufacturing provide sufficient solutions.

The results of inadequate PCS design become visible particularly in the form of high levels of work-in-process (WIP). Companies get into a vicious circle in which long and variable production lead times and high WIP levels reinforce each other. Besides the bound working capital, further negative consequences are induced and cause poor delivery performance, which is usually most harmful to the company, constrain flexibility, and hinder the implementation of short quality feedback loops. In addition, it is obvious that high labor or machine productivity can only be achieved with a suitable PCS that always ensures that the right amount of the right materials is in place for being processed.

---

[1] First mentioned by Pine (1993)
[2] First mentioned by Institute Iacocca (1991)

C. Karrer, *Engineering Production Control Strategies*, Management for Professionals, DOI 10.1007/978-3-642-24142-0_1, © Springer-Verlag Berlin Heidelberg 2012

Since the early 1980s, significant research efforts went into the area of PCS. However, it seems like the transfer of the concepts developed in research into industrial practice is not fully happening yet. There is a need for a practically applicable engineering approach to chose or design a PCS according to the needs of the company and its customers (Lödding 2008). Thereby, the three most prominent questions a PCS engineering framework needs to provide an integrated answer to are:

- To which level should WIP be limited to form a buffer between consecutive process steps?
- Where should the order penetration point (OPP) be positioned in the production flow?
- How should be dealt with the demand uncertainty upstream the OPP? Should a make-to-stock or make-to-forecast system be preferred?

The PCS engineering framework presented in this book takes an integrated and practice oriented view on these three questions. Thereby, special focus is put on the third question, which is frequently discussed in a dogmatic fashion by production planners and engineers. Besides decision support among a 'pull' driven make-to-stock (MTS) and a 'push' driven make-to-forecast (MTF) approach, a new hybrid approach is suggested. The hybrid approach enables a continuously variable combination of push and pull. Thus, the advantages of pull, like controlled inventories, and the advantages of push, like the use of forecast information, are integrated. It will be demonstrated that, under certain conditions, the hybrid approach significantly outperforms the pure strategies.

The framework addresses the complexity faced in practice and is in line with the philosophy of Lean Manufacturing. It will be applied to a real life case study from electronics manufacturing to demonstrate its practical applicability. Thus, the book provides the reader with a toolset and relevant knowledge to design a good production control strategy and to thus create the backbone of any competitively viable production system.

1.2   Suitable Industrial Context for Application

Depending on the continuity of its material flow, a production system can be characterized as either continuous or discrete. In continuous production, goods are constantly moved along a predefined routing. This type of production is usually found for flow-goods in the chemicals-, basic material-, or food industry. In discrete production, goods are moved between processes at certain points in time and mostly in batches (Hopp and Spearman 2008; Günther and Tempelmeier 2012; Schneeweiß 2002).

The PCS engineering framework presented in the following is designed for discrete single-piece flow or batch production systems. Moreover, the focus will be put on production systems that exhibit a minimum level of structural and dynamic complexity, so that the determination of a suitable PCS is not straight-forward and crucial to operational performance.[3] Thus, the PCS engineering framework is built to be applied especially in industries with the following characteristics.

Regarding structural complexity, production systems at which the PCS engineering framework is targeted at typically include a medium to high number of intermediate-/end-products within multistage serial or non-serial and potentially complex material flows. Moreover, usually there are significant changeover times and thus a need for batch production. Regarding dynamic complexity, processes are normally afflicted by variability due to disturbances or different standard processing times per process and per product. Exemplary disturbances that cause variable processing times and thus need to be considered in PCS engineering are breakdowns or speed losses. Also planned events like maintenance that are executed during production time need to be considered. From the market perspective, the production would characteristically face variable and potentially correlated customer demands together with unreliable forecasts. The customer lead time is possibly shorter than the production lead time so that no pure make-to-order (MTO) is possible.

Typical industry examples that usually fulfill a large portion of the characteristics above are electronics manufacturing (Printed Circuit Board (PCB) assembly) or press shops in the automotive industry.

The physical production system design is considered as fixed and the PCS is modeled based on the given design. A system redesign, like for instance the combination of separated processes into a flowline where possible, is not within the scope of the proposed framework. However, it is usually sensible to perform such activities before or in the course of changing the PCS.

1.3   Structure of the Book

The book is structured into seven chapters. After the introduction (Chap. 1), the fundamental concepts and coherences relevant for PCS design are presented and the following considerations are put into the context of current research in the field (Chap. 2). The literature review aims at giving a broad overview of current research in the field and explains contributions that motivated the following approach. Moreover, in Chap. 2, the relevance of the PCS for operational performance is established. Then, in Chap. 3, the PCS engineering framework is constructed, starting with an analysis of the relevant system design drivers and an approach to upfront reduce complexity. The generic model, whose parameters are later optimized to design a PCS, is formulated based on queuing network theory and its basic properties are explored mathematically. Integrated approaches to adequately

---

[3] For an introduction to the concept of complexity in the context of production systems, the interested reader is referred to Appendix 8.5

map production system-based and demand variability-based design drivers are proposed. Then, the subsequent Chap. 4 address how to optimize the parameters of the previously formulated generic model. Therefore, an objective function is formulated and a reusable simulation framework for numerical optimization is implemented. A process is designed to be able to efficiently search the solution space, which would be too large for pure brute force optimization. During this process, several opportunities for further reducing the complexity of the optimization are elicited. In Chap. 5, the upstream control part of the developed PCS engineering framework, which hosts the new hybrid MTF/MTS approach, is experimentally explored in order to find general decision rules and closed-form solutions for its control parameters. Moreover, the characteristics of the production system that drive the potential impact of the proposed hybrid MTF/MTS approach are investigated. The practical applicability of the PCS engineering framework is then demonstrated in a case study from electronics manufacturing in Chap. 6. In this case study, the new hybrid control approach is applied and its superior performance compared to the pure MTF or MTS strategies in combination with limited buffers between the planning segments is shown. The book concludes by summarizing the results, pointing out limitations, and giving hints for further applications and research in Chap. 7. The structure of the book is illustrated in Fig. 1.1.

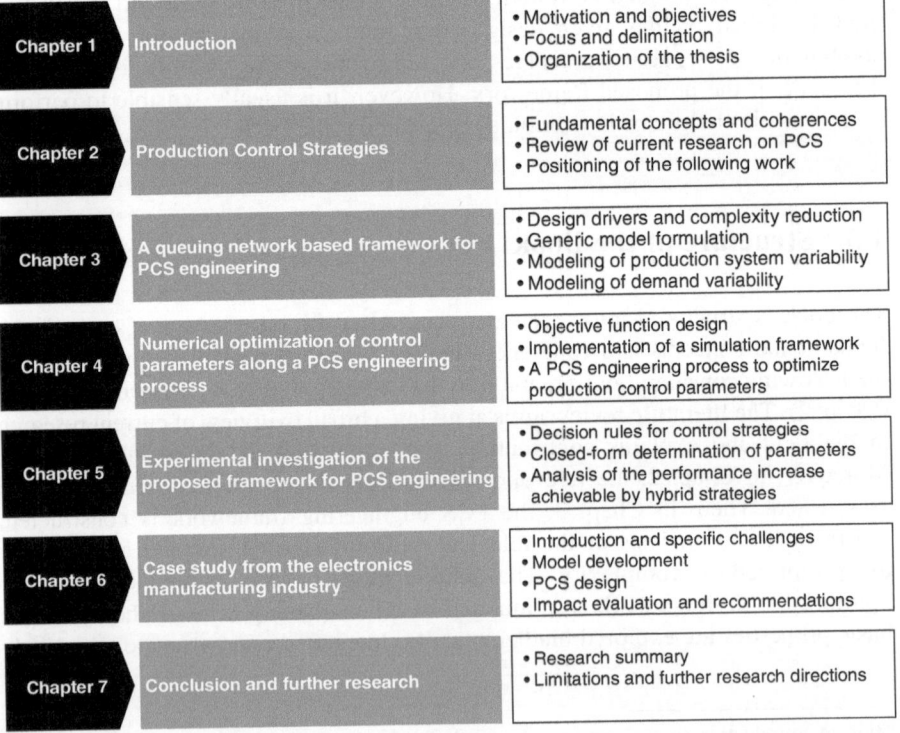

Fig. 1.1 Structure of the book

To make the following work best applicable in practice, it will go beyond pure mathematical modeling and take a systems engineering perspective. The quantitative methods used stem mainly from systems engineering, queuing network theory, decision analysis, and simulation. The work sticks closely to the thinking of Lean Manufacturing (Womack et al. 2007; Womack and Jones 2003) and assumes that the reader is familiar with its basic concepts.

Chapter 2
Production Control Strategies (PCS)

2.1 Fundamental Concepts and Coherences

2.1.1 PCS in the Broader Context of Production Planning and Control

The following discussions on PCS can be associated to the field of production systems engineering (PSE). "PSE is an emerging branch of Engineering intended to uncover fundamental properties of production systems and utilize them for analysis, continuous improvement, and design" (Li and Meerkov 2009). Other than classical manufacturing engineering, PSE is not concerned with the operation technology of machines or material handling devices, but with the parts flow through a production system (Li and Meerkov 2009). In this context, by production control strategy (PCS), it is referred to the information flow and the logic behind it that controls the movement of material within a factory (Hopp and Spearman 2008). Thus, it is focused rather on the information flow that controls the material flow, than on the actual physical material flow, which is the center point in the science of material flow (Arnold and Furmans 2009). In Fig. 2.1, the following work is positioned within the context of overall corporate operations planning. In the planning framework of Stadtler and Kilger (2008), PCS are part of the short term production planning.

Thereby, the focus of the presented PCS engineering framework lies on shopfloor control, or more specifically WIP control, the OPP allocation, and order release. Scheduling and lot-sizing problems,[1] which can also be relevant in short-term production planning, are not explicitly covered. The task of warehouse replenishment planning found in the short term distribution planning field can be seen as part of the following work. Neighboring planning tasks like personnel

---

[1] The interested reader is referred to Hopp and Spearman (2008)

C. Karrer, *Engineering Production Control Strategies*, Management for Professionals,   7
DOI 10.1007/978-3-642-24142-0_2, © Springer-Verlag Berlin Heidelberg 2012

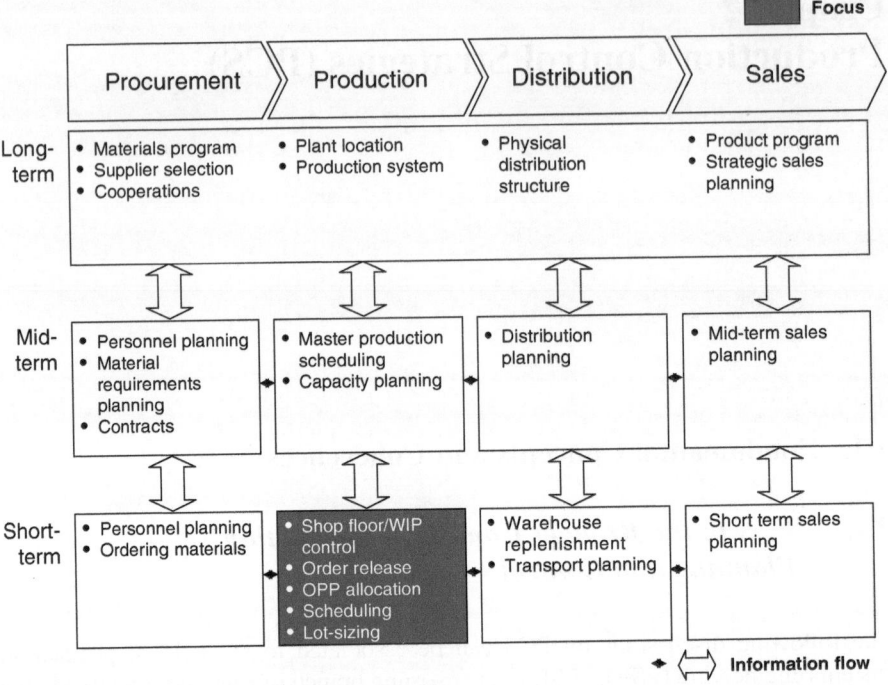

Fig. 2.1 Scope of PCS within overall planning (Adapted from Stadtler and Kilger 2008)

planning, capacity planning, outbound logistics planning, or the ordering of raw materials determine important operating conditions and thus parameters for PCS design.

2.1.2 The Push/Pull Enigma and Their Basic Implementations

PCS are often classified to be either of push- or pull-type. This distinction has caused lots of confusion and dissent among practitioners and researchers (Benton and Shin 1998; Bonney et al. 1999).

One reason for this is the large variety of often contradicting definitions used in literature and practice. Another reason for confusion originates from the fact that in practice, neither push nor pull are found in their purest form (Pyke and Cohen 1990).

The definition of Hopp and Spearman (2008) is found to be the most useful one to discuss PCS from a systems engineering point of view. According to them, the distinguishing feature is how the movement of work is triggered. In a push system, work orders are scheduled based on actual or forecasted demand by a central

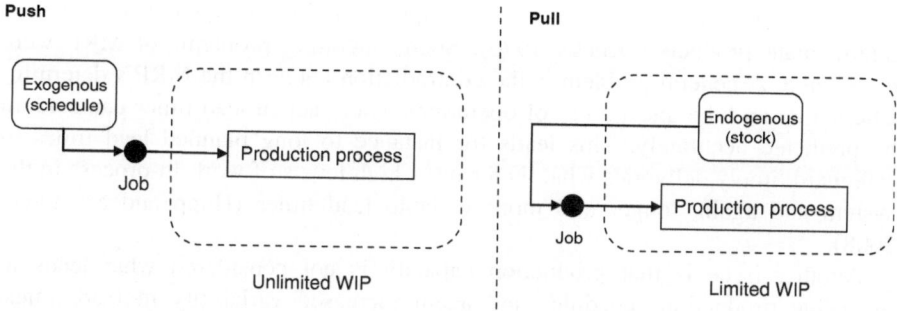

Fig. 2.2 Push and pull mechanics (Adapted from Hopp and Spearman 2008)

system. In a pull system, work is authorized based on the current system status. Figure 2.2 illustrates the two concepts.

Hopp and Spearman (2008) refine this definition based on the distinguishing effect of the two principles and state "A pull system establishes an a priori limit on work-in-process, while a push system does not." However, push systems are able to proactively operate based on forecasted demands what is not the case in pull systems, which only react to the actual status of the system. In practice, 'pull' is often used to refer to three different principles. The first two will be relevant in the following work. First, as described in the definition above, pull refers to the fact that WIP is limited between process steps, and a preceding process is only allowed to produce if sufficient space is available in the input buffer of the next process. Second, pull is used to describe a make-to-stock replenishment system, also called 'supermarket', in which different variants are stored to fulfill customer orders. Whenever a variant is removed from the supermarket, the same quantity is reordered and reproduced to fill up the empty spot. Third, with "pull" it is referred to a concept in internal production logistics, in which the production line is supplied with raw materials based on actual demands (Hopp and Spearman 2008; Womack et al. 2007).

Prominent implementations using mainly the push principle are material requirement planning (MRP) systems. Prominent implementations of the pull system are Kanban and Basestock. They will be briefly introduced in the following.

The idea of MRP systems was developed in the early 1960s by Joseph Orlicky (1975) as computer technology started to be used commonly by companies. The basic function of MRP is to calculate quantities and process start times for intermediate products (or ordering times for raw materials) based on actual or forecasted demands for final products. Thus, each process in the production system is planned and scheduled by a central system. The production orders are then 'pushed' into the system.

Due to the ability of MRP systems to process actual and forecast-based orders alike, MRP, and thus push, are often equalized with make-to-forecast (MTF). To translate the demand for final products into demands for raw materials and intermediate products, the so-called bill-of-material (BOM) is used. The BOM is a tree explaining on different levels the composition (type and quantity) of end and intermediate products (Orlicky 1975). Soon, operating problems of MRP were discovered. A general problem is the contradiction between the MRP's deterministic nature and the uncertainty of operations where actual lead times can seldom be predicted accurately. This leads for instance to long planned lead times to safeguard timely deliveries what then causes high levels of work-in-process in the system and again, longer and more variable lead times (Hopp and Spearman 2008).

Another issue is that production capacity is not considered what leads to infeasible production schedules and again increased variability in lead times. A further problem is called system nervousness and refers to the effect that small changes in the master production schedule can lead to large changes in planned order releases. Some of these problems could be mitigated by the introduction of manufacturing resources planning (MRP II), however, in shop floor reality, most of them remained (Hopp and Spearman 2008; Benton and Shin 1998). Today, MRP II software is usually part of comprehensive software packages called enterprise resource planning (ERP) (Jacobs and Bedoly 2003).

A popular pull-type implementation is Kanban. The Kanban system originates from the Toyota Production System (TPS) where it has been implemented as control mechanism for a production line in the mid-1970s. Kanban is Japanese for "card" and refers to the information carrier used to convey production authorizations between consecutive process steps (Ohno 1988; Kimura and Terada 1981). Even though a large variety of articles has been published on the topic, the definition of a Kanban system remains ambiguous. A summary of different definitions is provided by Berkley (1992). The mechanism is explained along the unified framework for pull control mechanisms developed by Liberopoulos and Dallery (2000) in Fig. 2.3. Each manufacturing stage MF_i has an input buffer I_i and an output buffer PA_i. Arriving Kanban cards are collected in queue DA_i.

Whenever a Kanban card is present in DA_{i+1} and the corresponding material is available in PA_i, the processing of this part type is initiated by launching it in the input buffer I_{i+1} of the next process. At the same time, the Kanban card is detached

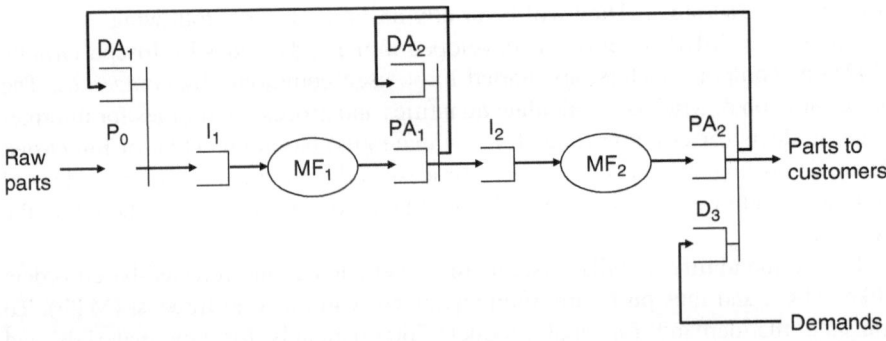

Fig. 2.3 Illustration of a two-stage Kanban system (Liberopoulos and Dallery 2000)

from the material and sent back to DA_i to reproduce the consumed part. The WIP in this system equals per definition the number of Kanban cards and is thus limited. The customer demand for end products is communicated stepwise upstream (Liberopoulos and Dallery 2000). It can be distinguished among Kanban systems that perform the blocking and production authorization per part type, and systems that block by total queue size (Berkley 1992). The first ones are in practice sometimes referred to as supermarket systems, the latter as sequential pull systems. The question how to set the number of Kanban cards has been extensively addressed in literature. A survey on this question can be found in Berkley (1992).

To close the introduction of the Kanban system, a few remarks about its relation to Just-in-Time (JIT) should be made. JIT stands for an approach that ensures that it is only produced what the customer needs, at the right point in time, in the right quantity, and with minimal lead time. To achieve this, JIT resorts to the tools of continuous flow, takt time, production leveling, and pull systems. Kanban is one option to implement a pull system. Thus, JIT is a superordinate concept to Kanban (Drew et al. 2004).

A similar way to implement the pull principle that even appeared earlier in literature than Kanban is Basestock (Clark and Scarf 1960). Applying the same framework and two-stage production system as used to illustrate the Kanban system, Basestock can be described as displayed in Fig. 2.4. In its initial state, the output queues P_i of the manufacturing processes MF_i contain a certain initial amount of stock, the so-called 'basestock' that gave the system its name. Whenever a demand event for an end-product occurs, it is instantly communicated to the demand queues D_i of all processes. Given that the needed inputs are available in P_{i-1}, production is started.

The distinctive feature when comparing Kanban and Basestock is that in Basestock, the demand information is immediately communicated to all processes, whereas in Kanban, is travels stepwise upwards against the material flow. The Basestock system is equivalent to the Hedging Point Control System (Liberopoulos and Dallery 2000).

A large variety of enhancements and combinations of MRP, Kanban, and Basestock were developed. An overview on them will be given later in this chapter in the course of the review of current PCS method design research.

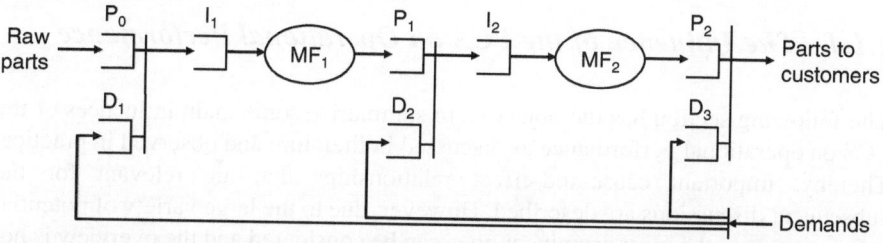

Fig. 2.4 Illustration of a two-stage Basestock system (Liberopoulos and Dallery 2000)

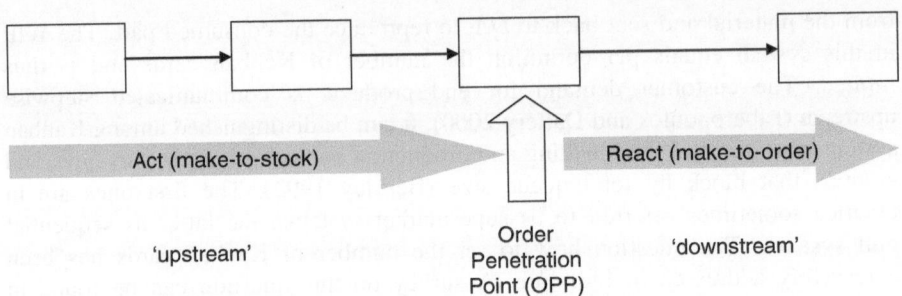

Fig. 2.5 The order penetration point (Adapted from Alicke 2005)

2.1.3   The Order Penetration Point

A concept that will be important within the PCS engineering framework constructed in the following is the order penetration point (OPP), sometimes also referred to as customer order point. "The order penetration point (OPP) defines the stage in the manufacturing value chain, where a particular product is linked to a specific customer order" (Olhager 2003). Figure 2.5 illustrates this idea.

To the processes left of the OPP it is referred to as 'upstream', to the processes to the right of the OPP as 'downstream'. For the upstream and downstream part, different production strategies have to be considered (Olhager 2003). Before the OPP, a way of dealing with uncertain demands needs to be found. After the OPP, a make-to-order system is feasible which does not have to hold any inventory to cover for uncertain demands. Therefore, moving the OPP as far upstream as possible, saves inventory (Alicke 2005). How far the OPP can be moved upstream depends on the comparison of the customer lead time, which is the time period the customer is willing to wait from order to delivery, with the production lead time, which is the time needed to complete the order from the potential OPP location to delivery. To determine the OPP location, also other criteria can play an important role. Examples include the customization options provided to the customers or the product structure in general. For a detailed discussion of these strategic considerations the reader is referred to Alicke (2005) or Olhager (2003).

2.1.4   The Influence of the PCS on Operational Performance

The following section has the objective to summarize some main influences of the PCS on operational performance as discussed in literature and observed in practice. Thereby, important cause-and-effect relationships that are relevant for the subsequent discussions are described. However, due to the large variety of potential influences, only the most prominent ones can be considered and the overview is not exhaustive. Operational performance is commonly measured in three dimensions:

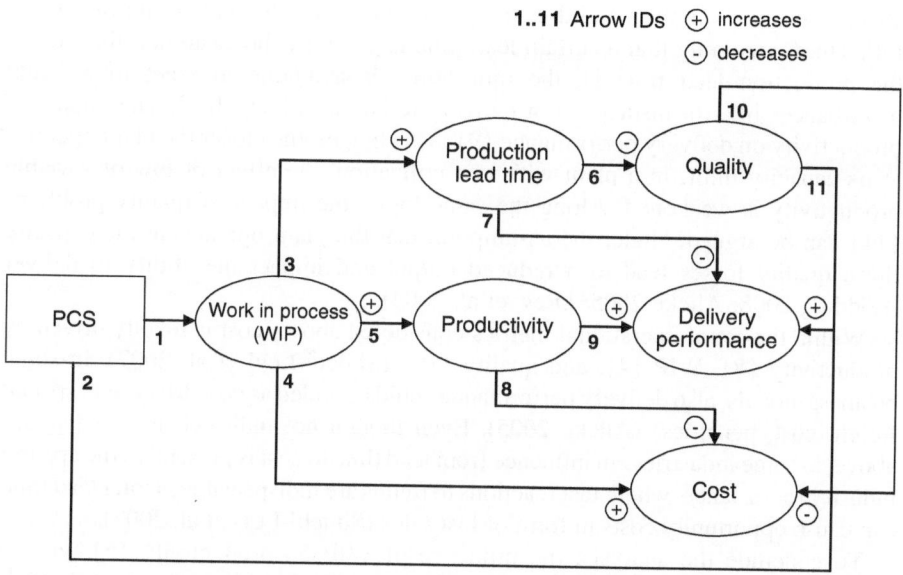

Fig. 2.6 Influence of the PCS on operational performance

quality, cost, and delivery performance (Drew et al. 2004). With the help of the illustration in Fig. 2.6, the most important causal chains from PCS over system characteristics to operational performance will be explained. On the arrows of the influence diagram it is indicated whether there is an influence that increases or decreases the impacted characteristic. Moreover, an identifier (ID) is assigned to each arrow to ease the following discussion.

Per definition, the PCS has a direct influence on WIP in the production system, i.e. its location, type, and amount {1}. The PCS has also an obvious direct influence on the delivery performance {2}. It needs to trigger material movements such that timely delivery is ensured (Hopp and Spearman 2008).

Besides the actual processing time, WIP is closely linked to production lead time, since it causes waiting time in front of processes {3} (Arnold and Furmans 2009).

The inversely proportional impact of production lead time on quality performance {6} can be explained with the idea of quality feedback loops. In many production systems, the quality of parts cannot be or is not directly assessed until a later process step. The longer the lead time to this process step is, the longer it takes until a potential error is discovered and solved. This also means that more parts are produced, potentially containing the error that would lead to scrap or rework (Drew et al. 2004).

Delivery performance is, besides the direct influence of the PCS, driven by the production lead time {7} and, under the assumptions of insufficient capacity, by productivity {9}. A longer production lead time has a negative influence on delivery performance for multiple reasons. First, the flexibility to react no short notice changes or orders is limited. Second, the longer the lead time is, the bigger is also its absolute variation and thus the probability to fail to deliver on time and in full. Third, assuming that a certain lead time is given by the customer, the shorter the production lead time is, the more time is available to react to external disturbances like for instance poor delivery reliability of suppliers. The impact of productivity on delivery performance {9} is the bigger, the closer the plant operates at its capacity limit. In a plant with low utilization, the effect of low or variable productivity is weakened. Along the same logic, the impact of quality problems {11} can be argued. Under the assumption that the plant operates at the capacity limit, quality losses lead to a reduced output and impact the ability to deliver (Lödding 2008; Alicke 2005; Drew et al. 2004).

Within the set of operational metrics considered above, cost is mainly driven by productivity {8}, WIP {4}, and quality {10} (Simchi-Levi et al. 2007). In some business models, also delivery performance could be added as cost driver (e.g. special freight cost, penalties) (Alicke 2005). Even though not indicated in the diagram above, in some industries, an influence from lead time to cost is present. In the apparel industry for instance, where fast reactions to trends are indispensible, a long lead time can cause opportunity costs in form of lost sales (Simchi-Levi et al. 2007).

To conclude the analysis, the influence of WIP on productivity {5} will be explained in more depth. In a production system with variable cycle times, buffers ensure a smooth operation by preventing processes from starving if the supplier does not deliver in time, or from blocking if the subsequent process is not ready to accept a new job. Thus, depending on the variability level, increasing WIP leads, under certain assumptions, to increased productivity. The productivity gain is diminishing with increasing WIP level. This basic coherence has been extensively explored by Nyhuis and Wiendahl (1999) within their operating curves approach (Fig. 2.7).

However, as mentioned above, this relation assumes that there are no other influences of WIP on productivity, which is not necessarily true. For instance, in space constraint environments, additional WIP can also reduce productivity by inducing waste (e.g. more motion required) (Drew et al. 2004).

Moreover, a further perspective can be added to this trade-off. From a time perspective, the improvement speed of the whole system, which is the speed in which variability and thus the need for WIP can be reduced, is the faster, the lower the WIP level is. On the one hand, low WIP levels derogate productivity, but on the other hand, make problems more visible. With low WIP levels, more processes are

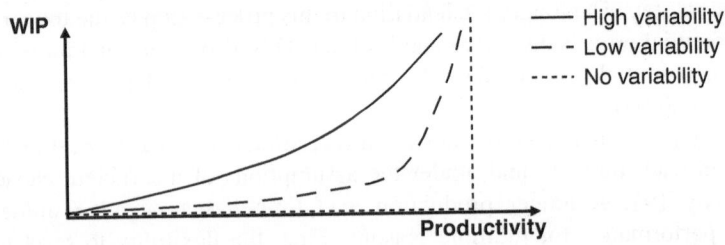

Fig. 2.7 Illustration of the trade-off between WIP and productivity (Nyhuis and Wiendahl 1999)

affected faster by problems and thus, more 'pain' is caused within the organization. This enables ultimately a more consequent root-cause-problem-solving of issues. This feedback loop is one of the core ideas in Lean Manufacturing (Womack et al. 2007).

2.2   Review of Current Research on PCS

2.2.1   Segmentation of Literature

In the following, an overview of current research in the field of production control strategies (PCS) will be given. Therefore, the existing literature is clustered and the main findings within each cluster are summarized and interpreted in the context of the following work. The literature review is fitted towards first, giving a broad and comprehensive overview on current research in the field, and second, towards introducing the groundwork on which the PCS engineering framework will be built on.

PCS related research is mainly addressed by publications from the field of Operations Research (OR) and production engineering. It is hard to overlook the vast body of literature. Thus, it is proposed to cluster the publications in the field according to a logical order from PCS method development, over PCS selection, to PCS implementation, and related design questions as depicted in Fig. 2.8. The discipline of PCS engineering, to which this investigation wants to contribute, spans across all four fields. Getting a broad overview is essential in order to develop a practically applicable and holistic PCS engineering framework.

Publications in the first area, PCS method development, create new, enhance existing, or help to parameterize PCS. Starting from basic push and pull approaches, a large variety of enhancements has been developed, also unifying characteristics of push and pull, leading to so-called hybrid systems. This cluster is well penetrated and can be considered as fundamental PCS research. The second field addresses the question of selecting an appropriate production control strategy for a given production system. Studies in this area compare the performance of selected PCS. The third cluster comprises implementation studies. Having chosen and customized the right PCS, implementation studies deal with how to turn these strategies into shopfloor reality. The fourth cluster bundles publications addressing design questions closely related to PCS design. Examples include scheduling, lot sizing, or inventory control. The literature review has its emphasis on the first three areas. Figure 2.9 shows a taxonomy of the relevant literature along which the remainder of this chapter will be structured.

The cluster of PCS method development can be split up further relying on the distinction between push and pull systems. The developed methods can be classified as either 'advanced push-type', 'advanced pull-type', or 'hybrid', which combine push and pull features. Hybrid systems can be distinguished further into 'horizontally integrated systems' and 'vertically integrated systems' (Cochran and Kaylani 2008). Vertically integrated systems consist of a higher-level push system

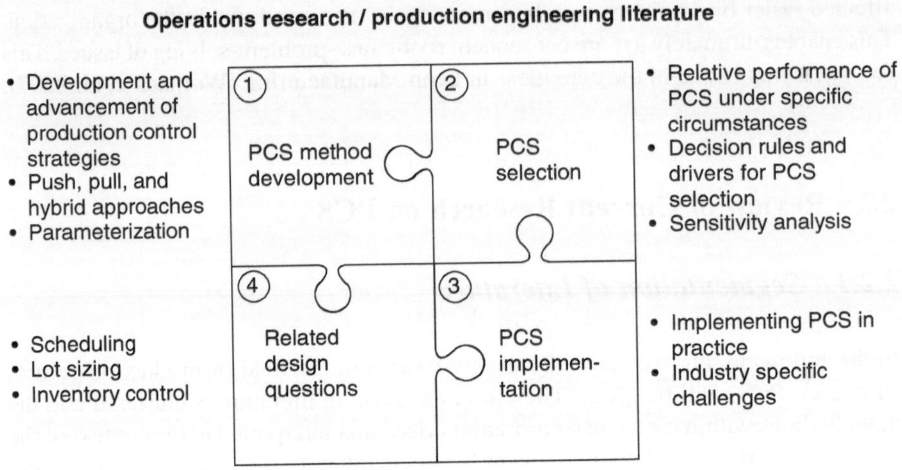

Fig. 2.8 Segmentation of PCS related literature

superimposed on a lower level pull system. In horizontally integrated hybrid strategies, some stages are controlled by the push principle and others by the pull principle. Figure 2.10 illustrates the two types of hybrid strategies.

Maes and VanWassenhove (1991) argue qualitatively that hybrid systems should be superior to pure push or pull systems in many application cases. In the following, important contributions to vertically and horizontally integrated hybrid systems, as well as to advanced pull- and advanced push systems are presented. The table in Appendix 8.6 gives an overview of the subsequently mentioned publications related to method design, thereby comparing their solution approach and major assumptions. A comprehensive survey of early publications on hybrid PCS is compiled by Benton and Shin (1998).

2.2.2 PCS Method Development

2.2.2.1 Vertically Integrated Hybrid PCS

One of the first contributions in this field comes from Hall (1986). He presents the "Synchro MRP" system used at Yamaha plants. Synchro MRP combines a classic MRP system with two card Kanban loops between all consecutive processes. Each process step is only authorized to produce, if an MRP production order and a Kanban card are present for the specific variant. Suri (1998) developed a similar system, the Polca (Paired-cell Overlapping Loops of Cards with Authorization) control system. In the Polca system, a central MRP system determines the start date of each production order in every process by backward scheduling. Polca cards rotate between two consecutive processes and authorize production. Only if the start date of a production order is reached, and a Polca card of the subsequent process

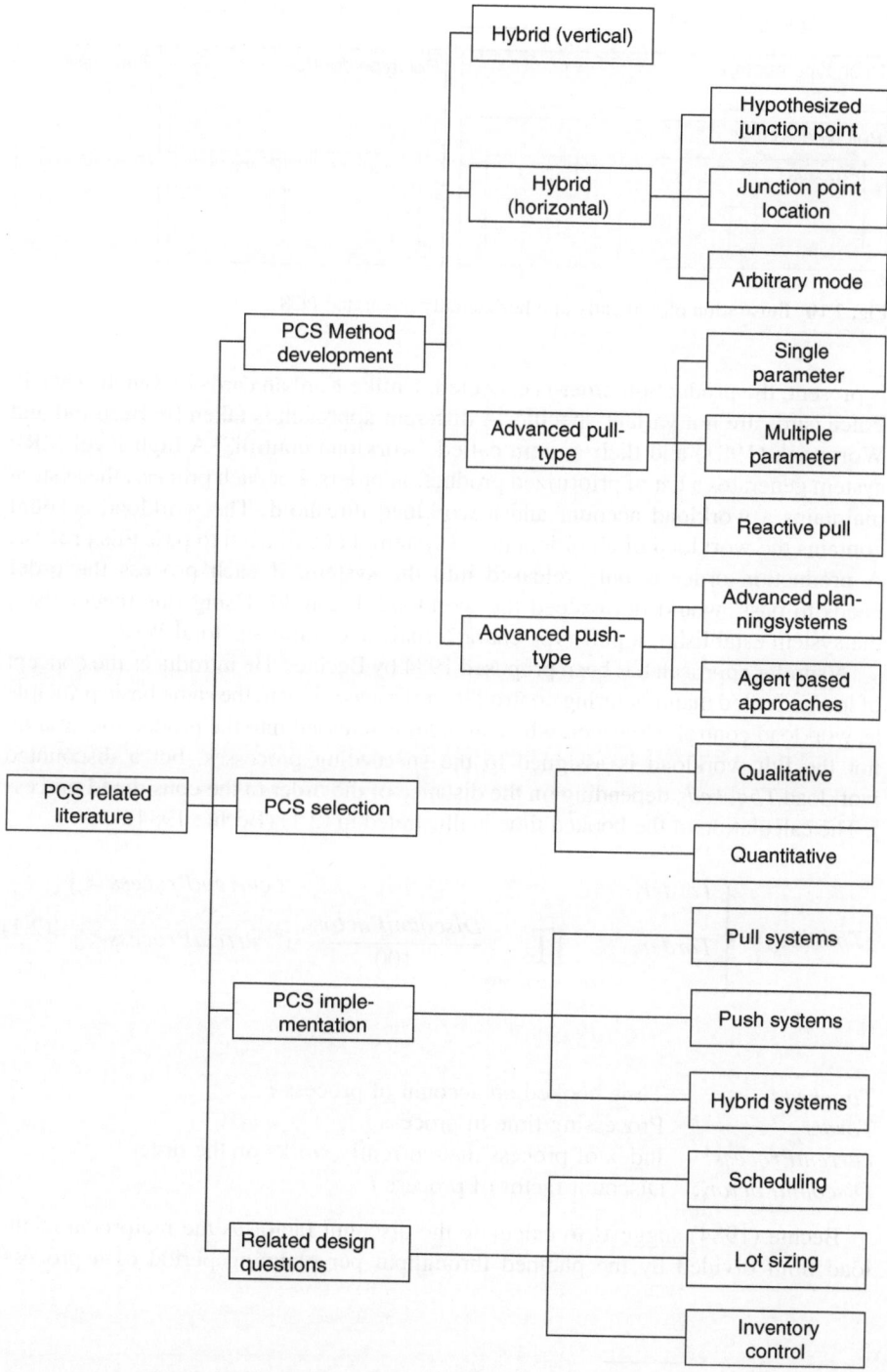

Fig. 2.9 Taxonomy of PCS related literature

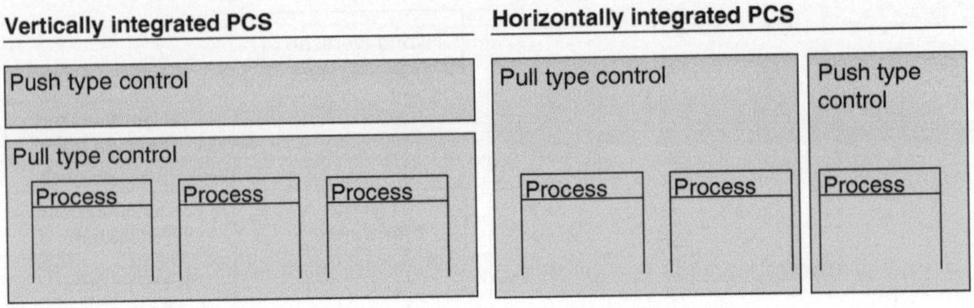

Vertically integrated PCS   **Horizontally integrated PCS**

Fig. 2.10 Illustration of vertically and horizontally integrated PCS

is present, the production order is executed. Unlike Kanban cards in Synchro MRP, Polca cards are not variant-specific. A different approach is taken by Bertrand and Wortmann (1981) and their system called "workload control." A high-level MRP system generates a list of prioritized production orders. For each process, the system maintains a workload account and a workload threshold. The workload account contains the workload of all orders in the system that still need to pass this process. A production order is only released into the system, if each process the order needs to pass, would not exceed the workload threshold. Using this mechanism, the system establishes a pull-type characteristic and limits the total WIP.

A similar approach has been proposed 1984 by Bechte.[2] He introduces the concept of load-oriented manufacturing control. It works according to the same basic principle as workload control. However, when an order is released into the production system, not the full workload is assigned to the succeeding processes, but a discounted workload $Tbooked_j$, depending on the distance of the order to the considered process j. The calculation of the booked time is illustrated in (2.1) (Bechte 1984).

$$Tbooked_j = \begin{cases} Torder_j & \text{if } currentProcess = j \\ Torder_j \cdot \prod_{i=currentProcess}^{j-1} \dfrac{DiscountFactor_i}{100} & \text{if } currentProcess < j \end{cases} \quad (2.1)$$

$Tbooked_j$:   Time booked on account of process j
$Torder_j$:   Processing time in process j
$currentProcess$:   Index of process that currently works on the order
$DiscountFactor_i$:   Discount factor of process i

Bechte (1984) suggests to calculate the discount factor as the reciprocal of the load limit divided by the planned throughput per planning period of a process.

---

[2] A description in English language of the concept can be found in (Bechte 1988)

This approach leverages the obvious coherence that the probability to actually complete a job within a planning period at a process is the bigger, the smaller the total waiting and processing time for a process is. This approach has been approved in practice but also been criticized for several reasons. The most important drawback occurs in production systems with low utilization. Here the completion probability of jobs is underestimated. An improved method in which the discount factor is independent from the load level is delivered by Perona and Portioli (1996).

2.2.2.2   Horizontally Integrated Hybrid PCS

Publications in this field address the question, which stages of a production system to control using the push principle, and which stages of a production system to control with the pull principle in order to create a hybrid system. The problem is solved either by modeling it as a Markov Decision Process (MDP) or by using discrete-event simulation. A distinctive feature of horizontal integration studies is the considered solution space. Along this criterion, three basic categories can be identified. In the first category, the location of a junction point[3] at which the control mode changes is directly hypothesized (for instance at the bottleneck). In the second category, the existence of one junction point is assumed and its location is determined via optimization. In the third category, each stage is allowed to either push or pull and the optimal control mode is determined for each stage via optimization.

A set of publications that directly hypothesizes the junction point location proposes to locate it at the bottleneck. Thereby, pull control is used from the bottleneck upstream and push control from the bottleneck downstream. This intuitive logic is used for example by the "Drum-Buffer-Rope" concept as described by Goldratt and Fox (1986), the "Starvation Avoidance" concept (Glassey and Resende 1988) or by the approach developed in Huang (2002). Beamon and Bermudo (2000) suggest a system that locates the junction point between sub and final assembly lines. Push logic is used for subassembly lines and pull logic within the final assembly line.

Olhager and Ostlund (1990) identify and describe further potential locations of the junction point. They propose to locate it according to the customer order point, the bottleneck, or the product structure. However, they do not provide guidance how to choose among the three options.

The problem of optimally locating the junction point and not hypothesizing it has been addressed by Takahashi and Soshiroda (1996) with the help of a set of difference equations. They allow the first processes to consistently either push or pull up to the junction point where the control mode alternates. They establish a relationship between the autocorrelation of the demand with the value of the

---

[3] Also known as 'push-pull-boundary' (Alicke 2005)

integration parameter. A similar problem is investigated by Hirakawa (1996) with means of simulation. Also Cochran and Kaylani (2008) picked up the junction point location problem. They focus on the question, whether each part type should have its own junction point or if a common junction point should be preferred. They minimize inventory holding and tardiness costs. Therefore, they optimize the junction point location, the safety stock level for the push stages, and the number of Kanban cards in the pull stages. The number of feasible solutions for a system with m stages, n parts, Q different counts of Kanban cards, and S different levels of safety stock equals according to Cochran and Kaylani (2008)

$$(S \cdot \sum_{i=0}^{m} Q^{m-i})^n \tag{2.2}$$

The underlying optimization problem is NP-hard and a genetic algorithm is applied to solve it. From simulation experiments and the application to a tube shop of an aerospace manufacturer, the following main conclusions are drawn:

• Horizontally integrated strategies can create value compared to pure push or pull strategies.
• If a bottleneck exists, the push-pull barrier should be located at the bottleneck process.
• Lower variability in parts arrival leads to lower safety stock.
• One junction point should be preferred compared to several product specific ones unless two parts sharing equipment have largely differing ratios of inventory holding cost to late cost.

Hodgson and Wang (1991a) studied the problem with a completely open solution space, e.g. each station can either push or pull. They developed an MDP for a four-stage iron and steel works production system. For the observed case example, they conclude that pushing in the first two stages and then pulling in the last two stages is a strategy with superior operational characteristics. They later extended their work (Wang and Hodgson 1992) to general parallel and/or serial multistage production systems. In the observed convergent material flow, they propose to push until the flows merge and to pull afterwards. Geraghty and Heavey (2004) later build on their model and show that in the way they modeled the pure push logic, it still has a WIP cap in each stage and thus their model is equivalent to a Kanban/CONWIP system, which will be presented as an advanced pull system later.

Hodgson and Wang's work has been extended by Pandy and Khokhajaikiat (1996) who introduce uncertain demand, production, and raw material supply. They then study a four-stage hair dryer production system. An overview of contributions to horizontally integrated hybrid systems can also be found in Geraghty and Heavey (2005).

2.2.2.3 Advanced Pull-Type Systems

In Sect. 2.1.1, the two most common pull-type production control strategies, Kanban and Basestock, were introduced. According to the pull definition of Hopp and Spearman (2008), they share the commonality that WIP is limited within them. In the following, extensions of these systems developed in current research will be presented. First, advanced pull-type systems that have one parameter per control loop are covered. Next, generic systems with multiple parameters per control loop will be investigated. The section concludes with what will be referred to as 'reactive pull-type systems' that adjust one or more of their parameters during operation to changing environmental conditions. To illustrate the advances in the field of one parameter pull-type systems, a three-step production line as displayed in Fig. 2.11 is used.

In Fig. 2.11, a classical Kanban system is displayed, which constraints the amount of WIP for each variant between two consecutive processes. A Kanban signal authorizes the reproduction of one unit as soon as an entity leaves the inventory. The single control parameter is the number of Kanbans per loop. This classic approach has been integrated with a variety of related optimization problems like the lot sizing problem (Li and Liu 2006). An overview of studies related to the classic Kanban system is provided by Berkley (1992).

In 1990, Spearman et al. introduced and studied the CONWIP (*Co*nstant-*W*ork-*In*-*P*rocess) pull system that puts a total WIP cap on the whole production line as displayed in Fig. 2.12.

Since its invention, the CONWIP system received lots of attention from research and got attested superior operational characteristics compared to other approaches by various studies. Framinan et al. (2003) provide an exhaustive survey of CONWIP related publications.

Bonvik and Gershwin (1996) combined the Kanban and CONWIP system as displayed in Fig. 2.13 to the Kanban/CONWIP system. Here, processes that are part of more than one Kanban loop, like the first process in Fig. 2.13, are only allowed to produce, if a card from each loop is present. It is shown that in a certain operating environment, this policy can achieve almost the same output with less WIP compared to pure Kanban or CONWIP systems. Kleijnen and Gaury (2003) attest this system a superior performance when robustness and risk are considered.

Finally, optimization models that allow installing arbitrary pull loops as displayed in Fig. 2.14 were developed. Within these systems, the challenge is to determine the optimal number of Kanban cards (allowed WIP) for each control loop. Control loops with an infinite number of cards are not implemented.

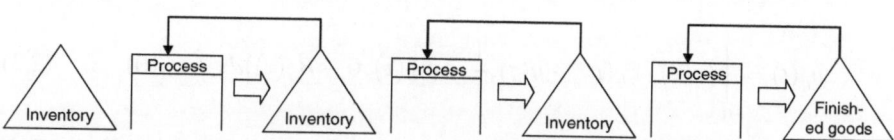

Fig. 2.11 Illustration of a Kanban system within a serial three-step production line

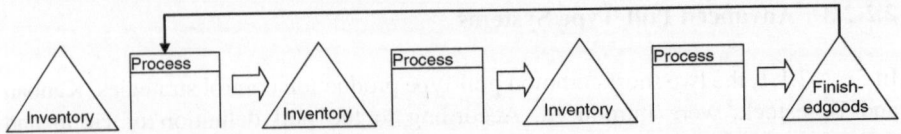

Fig. 2.12 Illustration of a CONWIP system

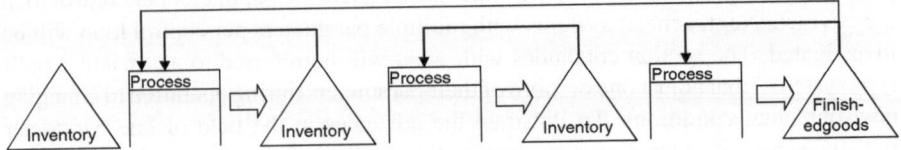

Fig. 2.13 Illustration of a combined Kanban/CONWIP system

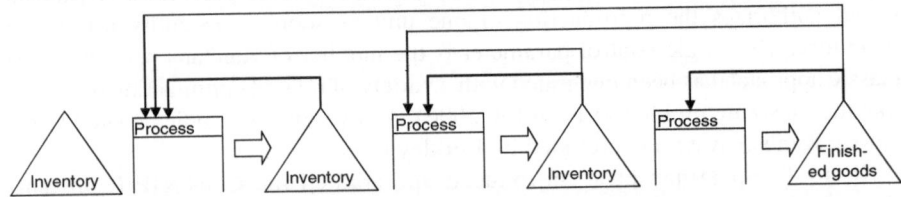

Fig. 2.14 Illustration of a production line with arbitrary pull loops

Gaury et al. (2001) apply discrete-event simulation and an evolutionary algorithm as heuristic to study system configurations. They perform a Plackett-Burman design (Montgomery 2009) with ten parameters varied on two levels and compare the resulting configurations. The observed parameters are displayed in Table 2.1 and include measures of the production system's structure, process variability, and demand variability.

They conclude that there is no dominant one-fits-all solution according to the considered WIP/delivery performance trade-off. However, they identify two important patterns that characterize most solutions. One links each stage to the first stage, and the other links the last stage to each preceding stage.

For the same problem, Masin and Prabhu (2009) apply a simulation-based feedback control algorithm called Average-Work-In-Process (AWIP) to derive solutions for the allowed WIP in each loop. In each simulation step, the number of Kanban cards in every loop is either increased or decreased. The number of cards $\mu_{ij}(t)$ at time t in the loop between stages i and j can be expressed as (Masin and Prabhu 2009)

$$\mu_{ij}(t) = \int_0^t k_{ij}(\tau) \cdot \xi_{ij}(\mu^* - \mu(\tau), \frac{1}{i} - w_{0i}(\tau), 0 - B_{ij}(\tau)) d\tau + u_{ij}(0) \qquad (2.3)$$

Table 2.1 Experimental design in Gaury et al. (2001)

Factor	Two factor-levels	
	+	−
Line length	4	8
Line imbalance	0	0.5
Imbalance pattern	Funnel	Reverse funnel
Processing time coefficient of variation	0.1	0.5
Machine reliability	Perfect	Exponential breakdown
Demand coefficient of variation	0	0.5
Demand rate/capacity	0.9	0.8
Service level target [%]	99	95
Inventory value ratio	1	2
Customers' attitude	Lost sales	Backorders

$\mu_{ij}(0)$ represents the initial number of Kanban cards. The integral accounts for the changes in each adaption round from 0 to t. $k_{ij}(t)$ is a gain function that determines the magnitude of the change. $\xi_{ij}(t)$ can take values -1 or 1 and thus determines whether the number of cards is increased or decreased. It is a function of actual and required times between departures of units from the process step, WIP, and blocking characteristics. A further discussion would go beyond the scope of this survey. By applying the feedback control algorithm above, Masin and Prabhu (2009) show that they can save up to 50% stock compared to classical Kanban systems.

In the pull systems discussed above, the demand information is propagated together with the Kanban cards in opposite direction of the material flow. Thereby, the number of Kanban cards is the only parameter of each control loop. The following systems have more than one parameter and allow for more sophisticated information flows. They are generic, meaning that depending on their configuration, they can emulate different PCS. The Extended Kanban Control System (EKCS) (Dallery and Liberopoulos 2000) and the Generalized Kanban Control System (GKCS) (Buzacott 1989) separate the demand information from the Kanbans. Both are a combination of the Kanban and the Basestock system. However, Dallery and Liberopoulos positioned the EKCS as an enhancement of the GKCS since it exhibits two advantages over it. First, the functions of the roles of the parameters are clearly separated, which enables easier optimization. Second, the demand information is propagated upstream faster in the EKCS (Liberopoulos and Dallery 2000).

The EKCS will be explained in more detail in the following. For an N stage serial production system with one product type, Fig. 2.15 illustrates the operation of the EKCS. Production on the manufacturing process MP_i is triggered if input material is available in queue PA_{i-1}, a Kanban is present in queue A_i, and a is demand present in queue D_i. $J_{i,j}$ denotes the synchronization station between processes i and j that triggers production if all prerequisites are met. The two parameters of each control loop are the number of Kanban cards K_i and the initial base stock level S_i. S_i Kanban cards are attached to the finished goods of MP_i and $(K_i - S_i)$ Kanban cards

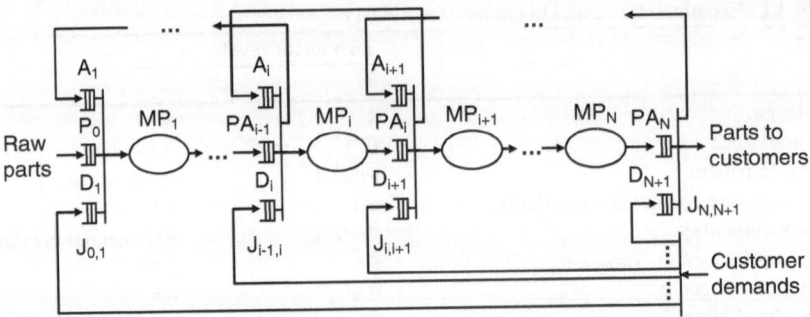

Fig. 2.15 Illustration of the EKCS (Dallery and Liberopoulos 2000)

are in queue A_i. Thus, during operation, the WIP level will be somewhere between S_i and K_i. The system therefore adapts its WIP level to varying demand and demand information is instantly passed on to all stages. It can be shown that the EKCS includes Kanban and Basestock as special cases (Dallery and Liberopoulos 2000).

An application and extension of the EKCS to non-serial flows in assembly manufacturing systems is demonstrated by Chaouiya et al. (2000). An extension to multiple products is provided by Baynat et al. (2002). Further possible enhancements of Kanban, EKCS, and the GKCS identified in Liberopoulos and Dallery (2000) are the introduction of WIP stage control, which limits the WIP of one single stage, and segmented systems, which nest different types of pull systems. Moreover, it is shown for several blocking mechanisms, like for instance minimal blocking (Mitra and Mitrani 1989) that they can be emulated by the approaches mentioned above. Comprehensive surveys of advanced pull-type systems can be found in Geraghty and Heavey (2005) and Liberopoulos and Dallery (2000).

A further generalization of the GKCS is the Production Authorization Cards (PAC) system (Buzacott and Shanthikumar 1992). The PAC system operates with a large variety of 'tags' among which the most important ones are production authorization cards, order tags, requisition tags, process tags, and material tags. Its basic operation will be explained using Fig. 2.16.

Production cells can request parts from stores using requisition tags. Process tags and order tags eventually form a production authorization (PA) card. The PA card is processed by the cell management. Thereby, order and requisition tags are generated and distributed to the preceding store. The cell management might introduce a delay between sending the process tag and the requisition tag. The operation of the cell management is not specified in detail in order to keep the system as flexible as possible. The PAC system is able to emulate a large variety of existing PCS approaches, among them Make-to-Order, Basestock, Kanban, MRP, and CONWIP (Buzacott and Shanthikumar 1992). The PAC system is a powerful approach but due to its high number of different tags, also a very complex system. The definitive paper on the PAC system is held very general in order to ensure a wide applicability and does not propose an approach for its customization. A comprehensive guideline for its customization has been delivered later by Rücker (2006).

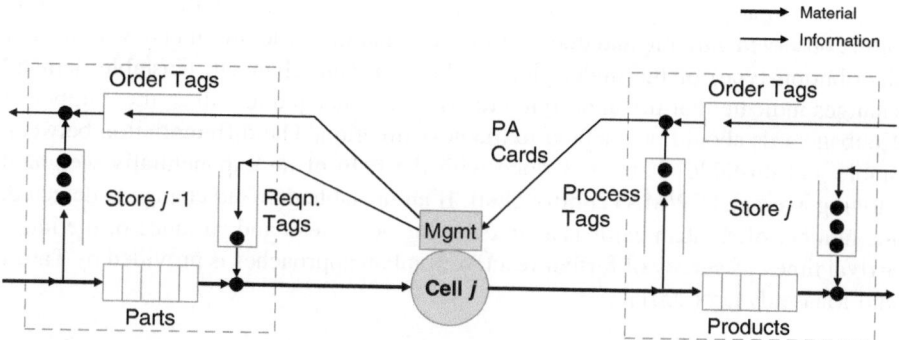

Fig. 2.16 Illustration of the PAC system (Buzacott and Shanthikumar 1992)

An interesting stream of publications examines the effect of advance demand information (ADI) on pull systems. Claudio and Krishnamurthy (2009) provide a survey on this field and examine the effect of perfect ADI on single and multiple product Kanban systems. The effect of imperfect ADI on a Basestock system in a single product single stage system is analyzed by Gayon et al. (2009). In their work, the ADI is imperfect in the sense that customers may cancel orders or order prior or later than expected. A similar scenario is analyzed by Liberopoulos (2008). He shows that in a basestock system, in which the ADI lead time is long enough, the basestock level drops to zero and that, under certain conditions, a linear relation between the basestock decrease and ADI lead time increase exists. Liberopoulos and Koukoumialos (2008) study the effects of variability and un-certainty on a Make-to-Stock supplier. Thereby, single items are ordered one at a time by two types of customers. The first type requires immediate delivery whereas the second type provides uncertain ADI in the form of cancellable reservations. The authors draw conclusions regarding the necessary stock in the capacitated and uncapacitated case. Babai et al. (2009) suggest a new dynamic reorder point policy that considers externally given and known forecast uncertainties in a single-stage single-product production system. Benjaafar et al. (2010) examine a single-product production system with stochastic processing times. ADI is provided and updated with variable lead time. Orders can be cancelled. They suggest for future research to "consider systems where order sizes are variable and where the actual number of units in each order is not exactly known until the order becomes due", an aspect picked up in the model developed in the next section. Further work examining the impact of ADI on PCS has been provided by Tan et al. (2007), Karaesmen et al. (2004), and Liberopoulos et al. (2003). The groundwork for integrating ADI in Basestock systems can be found in Karaesmen et al. (2002).

A reactive Kanban system has been proposed by Takahashi and Nakamura (2002) and will be described in the following. The demand and its variability is one of the drivers for the number of Kanban cards within a loop. Therefore, they monitor the inter-arrival time of customer orders in order to detect the need to recalculate the number of Kanban cards. It is distinguished between stable and unstable demand changes. A stable demand change comes from the random nature of the observed variable and does not indicate that the structure (mean, variance, or distribution type) of the underlying random variable changed. Unstable demand changes indicate that the underlying distribution changed and thus, the number of Kanban cards should be adapted to the new situation. The differentiation between stable and unstable changes is done with the help of an exponentially weighted moving average (EWMA) control chart. If an unstable demand change is detected, the number of Kanban cards is adapted to the new mean and variance of the inter-arrival times. A survey of further reactive Kanban approaches is provided by Tardif and Maaseidvaag (2001).

2.2.2.4 Advanced Push-Type Systems

To overcome the weaknesses of classical MRP push systems, they were not only combined with pull systems, but also efforts to improve the underlying push logic itself were made. The integration of algorithms from Operations Research, mostly heuristics, lead to so-called Advanced Planning Systems (APS) (Günther and Tempelmeier 2012). They still rely on the basic centralized push logic, but are able to solve more complex planning or sequencing problems. A detailed survey on the extensive amount of literature available on APS will not be given here. The interested reader is referred to Günther and Tempelmeier (2012), Stadtler and Kilger (2008), or Tempelmeier (2001).

During the last decade, also agent-based production control systems were investigated, for instance by Gelbke (2008), Mönch (2006), and Khoo et al. (2001). Mönch (2006) proposes a framework for a distributed hierarchical control system for the semiconductor industry, called FABMAS (Fab-multi agent system), whose high level architecture is shown in Fig. 2.17. FABMAS performs planning on three levels: productions system, production area, and process group.

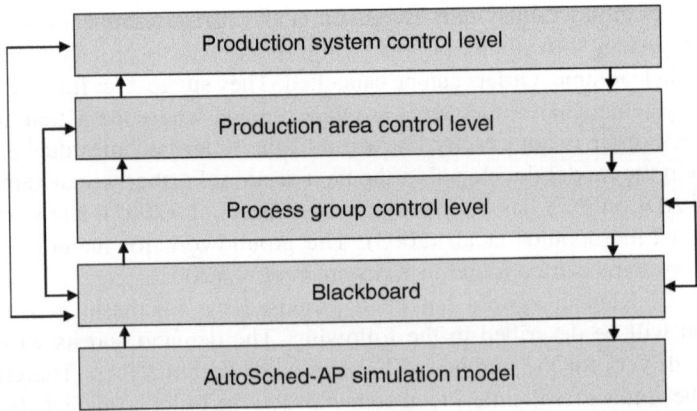

Fig. 2.17 High-level architecture of the FABMAS prototype (Translated from Mönch 2006)

The blackboard represents a data layer that is used by agents to exchange information. The bottom layer is a discrete-event simulation model used to test the system. All entities in the production system are modeled as agents (e.g. product agents, batch agents, service agents, decision maker agents, and so on). The three planning levels are supplemented with heuristics that resort to central information. The production system control level uses for instance a beam-search-algorithm, the production area control level a distributed-shifting-bottleneck-heuristic, and the process group control level relies on a machine hours-based resource allocation algorithm. These algorithms are also executed by designated agents. Production plans evolve through the performed optimizations and interactions among agents (Mönch 2006).

Agent-based approaches are promising due to their decentral, flexible, and complexity reducing character. However, up to today, they are embedded in systems following the push logic. It still needs to be investigated how they perform facing the typical pitfalls of push/MRP systems in practice. An integration with pull systems, for instance by using an agent-based approach for scheduling, could be worth investigating.

Comprehensive surveys of earlier PCS design literature are given by Geraghty and Heavey (2005), or Benton and Shin (1998).

2.2.3 PCS Selection

The following section focuses on publications that aims at deriving information on the relative performance of PCS, thus leading to drivers and decision rules for PCS selection. In the presented studies, usually two or more PCS are placed into a hypothetical production system and their performance is evaluated according to one or more metrics. In some cases, also the influence of environmental factors (e.g. presence of emergency orders) and characteristics of the production system are analyzed. Most studies argue on a quantitative basis by modeling the system with the help of Markov Decision Processes (MDPs), Petri nets, or discrete-event simulation models. Even though also qualitative studies (e.g. Razmi et al. 1996; Razmi et al. 1998) exist, the focus will be on quantitative studies here.

In the following, an analysis of 21 recent PCS comparison studies identified during a literature review is provided. Further surveys of PCS comparison literature are contained in Geraghty and Heavey (2005) and Benton and Shin (1998). The following dimensions are considered and represent the column headings in Table 2.2:

- *Compared PCS,* denotes the solution space, lists the PCS considered in the study
- *Factor variation,* denotes the factors whose impact on the PCS performance and decision were analyzed
- *Performance comparison,* describes how the performance comparison is made (metrics, approach, production setting)
- *Conclusion,* summarizes the conclusions drawn from the study

Table 2.2 Analysis of PCS selection studies

Reference	Compared PCS	Factor variation	Performance comparison		Production setting	Conclusion
			Metrics	Approach		
Bonvik and Gershwin (1996)	Kanban, Minimal Blocking, Basestock, Kanban/CONWIP	N/A	WIP, delivery performance	Discrete-event simulation (C++ implementation)	Serial production line, single part type, four stations	Hybrid dominated followed by CONWIP and then basestock
Geraghty and Heavey (2005)	Kanban/CONWIP, Kanban Hybrid (horizontal), Basestock, GKCS, EKCS	N/A	WIP, delivery performance	Discrete-event simulation (EM Plant)	Five stage parallel/serial line, one product produced out of two subassemblies, random process failures	Kanban consistently the worst performer, Kanban/hybrid consistently best performer
Grosfeld-Nir et al. (2000)	Push, Pull	Uncertainty in processing times, number of stations	WIP, throughput	Discrete-event simulation (SIMAN)	Serial production with 1–20 stations and stochastic processing times	If the number of stations is larger than seven, push dominates else pull
Gstettner and Kuhn (1996)	Kanban, CONWIP	N/A	WIP, throughput	Discrete-event simulation	One product serial flow line, exponentially distributed processing times	Kanban dominates
Hoshino (1996)	Push, Pull	Demand variation, variation of forecast error	WIP	Analytical approach	Single process step examined	Push dominates if variation of forecast error is small relative to variation of demand
Huang et al. (1998)	MRP, Kanban, CONWIP	N/A	WIP, throughput, raw material consumption rate, machine utilization	Discrete-event simulation	Cold rolling plant, six stages, convergent and divergent flows	CONWIP dominates
Kilsun et al. (2002)	Push, Pull (Kanban)	Demand variation, emergency orders	Cost (WIP holding and setup cost)	Discrete-event simulation (WITNESS 7.0)	Two stages serial production, ten products, only demand uncertainty	With low demand variation and no emergency orders push dominates, else pull dominates
Kleijnen and Gaury (2003)	Kanban, CONWIP, Kanban/CONWIP, Generic Kanban (see Gaury et al. 2000)	N/A	WIP, short-term delivery performance	Discrete-event simulation, Monte Carlo simulation	Serial production line, single product, four stations	Hybrid system dominates

(continued)

Table 2.2 (continued)

Reference	Compared PCS	Factor variation	Performance comparison		Production setting	Conclusion
			Metrics	Approach		
Koh and Bulfin (2004)	CONWIP, DBR (Drum-buffer-rope, horizontally integrated hybrid system with junction point at bottleneck)	N/A	WIP, throughput	Markov process model	Serial production line, three stations, unbalanced, exponential processing times	DBR dominates
Lee and Lee (2003)	Push, Pull	N/A	WIP, throughput	Discrete-event simulation	TFT-LCD production facility	Pull operates with much less WIP at the price of a minor decrease in production
Ozbayrak et al. (2004)	Push, Pull (Kanban)	N/A	Activity based costing	Discrete-event simulation (SIMAN)	MTO system (6 month demand freeze), six cells, 35 part types for ten products, scrap, rework, breakdowns	Pull dominates
Ozbayrak et al. (2006)	Push, Kanban, CONWIP	N/A	WIP, delivery performance, mean flow time, responsiveness	Discrete-event simulation	10 components assembled into one final product, different routing per component, processing times normally distributed	Depending on the metric, either push, Kanban or CONWIP dominated
Papadopoulou and Mousavit (2007)	Push, CONWIP	Dispatching rule (first-come-first-served, shortest-imminent-processing-time, earliest-due-date, work-content-in-the-queue-of-next-operation)	Average WIP, mean flow time, deviation from due date (earliness, tardiness), total time in queue	Agent-based simulation	Job shop, eight workstations, ten job types, revisiting of processes possible, batch size of 10–50 units	CONWIP dominated push under most dispatching rules and in most metrics
Persentili and Alptekin (2000)	Push, Pull	Product flexibility	WIP, throughput, average flow time, backorder level	not specified	Two products, five stages, convergent and divergent material flows	No clear dominance found
Razmi et al. (1996) and (1998)	Push, Pull (Kanban), hybrid (not specified further)	Variety of cost, flexibility and market factors	Utility function	Analytical hierarchical process	N/A	N/A

(continued)

Table 2.2 (continued)

Reference	Compared PCS	Factor variation	Performance comparison		Production setting	Conclusion
			Metrics	Approach		
Sarker and Fitzsimmons (1989)	Push, Pull (Kanban)	Cycle time variation	WIP, throughput	Discrete-event simulation	Three stations, serial flow, one product, exponentially distributes process times	The higher the coefficient of variation of the processing time is, the better performs push compared to pull
Sharma and Agrawal (2009)	Kanban, CONWIP, Hybrid	Demand distribution	Utility function	Analytical hierarchical process	Multistage serial production, single part	Kanban dominates
Tsubone et al. (1999)	Push, Pull (Kanban)	Breakdowns, processing time, flow sequence	Lead time	Discrete-event simulation	Serial flow, different jobs with different process sequences	Only small differences between push and pull, however push is more sensitive to internal variations
Wang and Xu (1997)	Push, Pull (Kanban), Hybrid (horizontal, each stage can either push or pull)	N/A	Inventory, delivery performance	Discrete-event simulation	Several production systems investigated from linear to an arbitrary convergent material flow, random demand and processing capacities	Hybrid dominates
Weitzman and Rabinowitz (2003)	Push, Pull (CONWIP)	Information updating rate, failure characteristics of machines	WIP, delivery performance	Discrete-event simulation	Serial flowshop, eight machines	Pull dominates, the worse the information updating the better performs pull

The objective of the analysis shown in Table 2.2 is to give an overview of the state of the art and to identify commonalities in the studies' conclusions, but also their weaknesses with regard to practical applicability.

The different conclusions drawn by the individual studies indicate that no per se dominant solution exists. The preferred approaches differ on a case by case basis and seem to be sensitive to changes in the assumptions or the production setup. However, it looks like that in most cases, especially if uncertainty is accounted for in the underlying models, pull-type or hybrid systems dominate pure MRP systems. Bonney et al. (1999) raise the question, whether the frequently argued superiority of pull-type systems compared to push-type systems is rather caused by their prerequisites, like small batch sizes, than by the different control of the material flow. Spearman and Zazanis (1992) suggest that the effectiveness of pull systems (i.e. less congestions) does not result from "pulling," but from the effect of limiting WIP and thus the variability of the amount of WIP. Moreover, they argue that in practice, pull systems are inherently easier to control since they focus on controlling WIP, other than push systems, which focus on controlling throughput, which cannot be visually observed like WIP. By examining the attributes in the columns of Table 2.2, further observations can be made.

Examining the applied factor variations, it can be concluded that the factor variation is seldom analyzed and thus the actual drivers for a decision remain unknown. However, this could be of interest especially for practitioners since they would know, which factors to monitor in order to get a signal to adapt the applied PCS to a constantly changing environment. The analyzed studies judge the PCS performance according to different metrics. WIP level and delivery performance are however frequently used metrics. To provide decision support in practice, the different objective functions and risk attitudes of the companies and their managers need to be taken into account. Trade-offs between metrics should be part of a utility function used for performance evaluation. Looking at the production systems in which the PCS are compared, the studies usually resort to very simplistic production setups with mostly serial flows, few products, and sometimes even deterministic parameters (e.g. cycle times, demands). Variability in the production system and within demand and forecasts is mostly addressed very rudimentary. Still, this helps to provide a directive value. Hence, the influence of complexity on the PCS decision could be explored in more depth.

An interesting approach is taken by Kleijnen and Gaury (2003). They apply a risk analysis-based approach in order to judge the robustness of the system. They combine discrete-event simulation, heuristic optimization, risk analysis, and bootstrapping and come up with a different ranking of the systems compared to a study (Gaury 2000), in which risk is not considered.

Geraghty and Heavey (2005) perform an exhaustive study in which they compare Kanban, Basestock, Generalized Kanban Control Strategy (GKCS), Extended Kanban Control Strategy (EKCS), and the horizontally integrated hybrid PCS presented in Hodgson and Wang (1991A), which equals the Kanban/CONWIP advanced pull system according to Geraghty and Heavey (2004). They conclude that Kanban/CONWIP consistently dominates all other PCS, and that Kanban consistently performs the worst. The dominance compared to Kanban is significant, the dominance compared to the other approaches is rather small, especially in the case of low system utilization. It is argued that the dominance of Kanban/CONWIP and CONWIP is due to the direct communication of demand information to the first stage that is inherent the CONWIP system and not done by pure Kanban, where the demand information is stepwise propagated upstream the material flow. However, the positive side of this information delay is not pointed out. It can also be interpreted as option to decrease the customer lead time and react to demand uncertainty since inventory of finished goods or semi-finished that are not yet dedicated to customer orders are held. A similar performance like in the Kanban/CONWIP system is shown by the EKCS. The main difference is that through its base stock levels, the EKCS tends to hold more of the inventory within the line whereas the Kanban/CONWIP systems tends to hold more inventory as finished goods. The details of these studies are presented in Geraghty (2003).

2.2.4 PCS Implementation

After choosing and customizing an appropriate production control strategy in theory, the next challenge is to implement it sustainably on the shopfloor. Therefore, several aspects need to be addressed appropriately to enable a sustainable implementation. They comprehend suitable

- Hardware
- IT systems
- Organization and processes
- Mindsets of involved employees

Only if all aspects above are addressed and potential industry specific challenges are solved, a sustainable implementation is possible. The available literature in this field can be categorized into publications that treat the implementation of pull-systems, publications dealing with the implementation of push systems, and publications dealing with hybrid systems.

Implementation studies in the first category, pull systems, focus mainly on the Kanban system. They are found in practice-oriented Lean Manufacturing publications like Womack et al. (2007), Drew et al. (2004), Womack and Jones (2003), or Ohno (1988). They treat the problem as part of holistic Lean transformations. Lee-Mortimer (2008) provides an implementation study of a Kanban system at an electronic manufacturer and stresses the aspects of the IT implementation, but also cultural aspects. Schwarzendahl (1996) describes the introduction of Kanban in a Siemens plant producing digital electronic switching equipment. One key challenge encountered in almost every implementation of a pull system is how to integrate the new control logic in the existing MRP system of the company. Flapper et al. (1991) present a three-step solution approach for this issue. A comprehensive survey of further IT-focused implementation studies can be found in Benton and Shin (1998). Slomp et al. (2009)

demonstrate in a case study how CONWIP control can be implemented together with other Lean Manufacturing tools in high-variety/low-volume make-to-order environments.

In the second category, push systems, literature focuses mostly on IT implementations of MRP systems or APS. Here, a large body of literature exists in the area of business computing. It will not be surveyed in detail here. A good introduction to software implementations of MRP systems is given by Scheer (2007) and for APS by Stadtler and Kilger (2008).

Some publications also study the implementation of hybrid PCS, the third category. Wang and Chen (1996) suggest an IT implementation of a horizontally integrated PCS by integrating a pull mechanism into the master production scheduling of MRP II. Goncalves (2000) studies the horizontally integrated hybrid PCS used in Intel's microprocessor fabrication facilities with the help of a system dynamics model. His focus is determining the influence of endogenous customer demand. Therefore, he assumes that the inventory levels influence demand. He models the behavior of several stakeholders (e.g. planners, warehouse managers, marketing) and studies the company's resulting production and service levels. Thereby, also insights on the behavior of the system in presence of failure (e.g. stock outs) are gained. Besides classical simulation analysis, he applies eigenvector analysis and derives stabilizing policies. Another example for a hybrid implementation study comes from Xiong and Nyberg (2000) who describe the implementation of a horizontally integrated hybrid system in refineries. The implementation of a push/pull production system for a single product production system in the food industry is described in Claudio et al. (2007). Here, different strategies are applied for different customers, depending on whether the customer provides forecast information or not.

2.2.5 Related Design Questions

Scheduling, lot-sizing, and inventory control are further design questions from the field of production planning and control, which are closely linked to the development of PCS. To conclude the literature review, they will be briefly addressed.

Scheduling is a discipline mainly addressed within Operations Research and computer science. "It deals with the allocation of resources to tasks over given time periods and its goal is to optimize one or more objectives" (Pinedo 2008). In the context of PCS, scheduling determines, for a process and its preceding buffer, which job to process next. It addresses thus an integral design question of every PCS and may influence its operational performance.

It is distinguished between single machine or multi-machine problems, in which schedules for several parallel processes are derived. Moreover, it is differentiated between flow-shop and job-shop problems. In a flow-shop, all goods flow along the same route through production (e.g. steel milling). In a job shop, goods follow, usually in batches, different routes (Pinedo 2008).

In practice, either no rules or standards exist, or very simple dispatching rules are applied. Common dispatching rules are First-In-First-Out (FIFO), Earliest Due Date (EDD), Shortest-Processing-Time (SPT), Shortest-Remaining-Processing-Time (SRPT), Longest-Processing-Time (LPT), Longest-Remaining-Processing-Time (LRPT) (Günther and Tempelmeier 2012; Framinan et al. 2000). An investigation of the impact of different dispatching rules on a CONWIP flow-shop is presented in Framinan et al. (2000). In their experiment, SPT outperformed FIFO and SRPT. Also Vinod and Sridharan (2009) present a simulation analysis of different scheduling rules. They investigate them in a job shop and consider sequence dependent setup times. They find the setup oriented version of SPT to perform best.

Another prominent approach to scheduling is earliness-tardiness scheduling. Here, the objective is to dispatch the orders in a sequence such that the deviation from the planned completion date is minimized (Baker and Scudder 1990). Lately, also optimization approaches from the area of artificial intelligence became popular in solving scheduling problems. Jensen (2001) presents an approach to solve stochastic scheduling problems in order to create robust and flexible systems with the help of evolutionary algorithms. Framinan et al. (2001) present a performance comparison of different heuristics. A high-level summary of the field can be found in Stuber (1998). However, also the argument exists that more complex scheduling rules, or the combination of simple scheduling rules, cannot significantly improve the performance of a production system, but rather generates a vicious circle by continuously reprioritizing production orders (Sabuncuoglu and Bayiz 2000; Stuber 1998; Kosturiak and Gregor 1995). For a general introduction to scheduling, the reader is referred to Pinedo (2008), Brucker (2007), or Blazewicz et al. (2001).

A second PCS design related area is lot-sizing. However, it will not be treated in detail here. The interested reader is referred to Günther and Tempelmeier (2012) or Hopp and Spearman (2008). Other than for instance Li and Liu (2006), this work does not integrate lot sizing in the optimization model but assume that optimal batch sizes are in place.

A third area closely related to PCS design questions is inventory control. "Inventory control determines which quantity of a product should be ordered when to achieve some objective, such as minimizing cost" (Kleinau and Thonemann 2004). Inventory control problems can be categorized depending on whether their demand is stationary, and whether it is fully observed (Treharne and Sox 2002). Inventory control can be delimited from the field of production control by the scope. Inventory control focuses on one single inventory, often of finished goods. In contrast, the scope of production control is broader. Production control looks at the whole production system and is in addition concerned with question where to set up inventories and the overall material and information flow in the production system. However, the terms are not consistently used in literature. A good survey on inventory control approaches is provided by Babai et al. (2009).

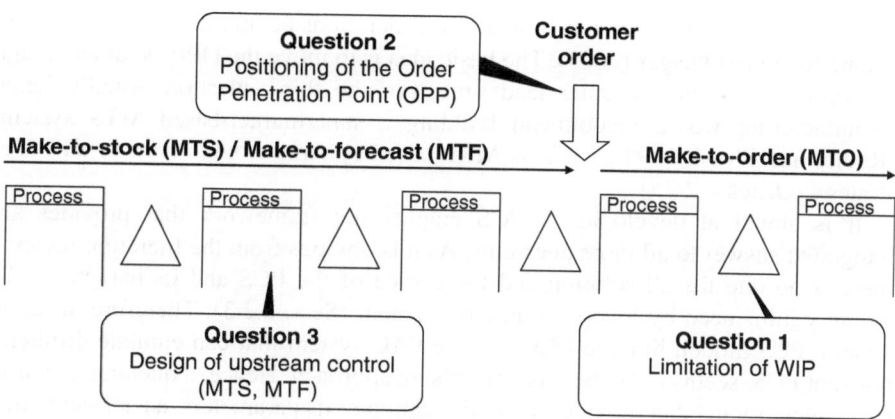

Fig. 2.18 Key questions of PCS engineering in practice

2.3   Synthesis and Positioning of the Following Work

After understanding basic concepts of PCS, their influence on operational performance, and current research in the field, now conclusions will be drawn and the
following work be positioned.

The analysis of the PCS selection literature (Sect. 2.2.3) with regard to the
creation of a PCS engineering framework shows that currently, a systematic and
comprehensive guide to engineer a PCS for a complex discrete real world production system is missing. This need has also been recognized by Lödding (2008) who
surveys different PCS but notes that further research is necessary to decide among
them. Existing comparison studies with strong assumptions and limited solution
spaces can provide directions, but are not able to provide concrete answers for
complex real world production systems. The following PCS engineering framework
seeks to contribute to closing this gap. Thereby, it is essential to provide an
integrated answer to the three most important questions of PCS engineering as
asked by industrial practice. Inspired by a Lean Manufacturing view on PCS, they
can be structured as shown in Fig. 2.18 (Rother and Shook 1998).

On the first question, a large amount of literature has been published as outlined
in the review of PCS method development literature (Sect. 2.2.2.3). Most papers
conclude that limiting WIP in the buffers is beneficial, in other words that pull
systems dominate pure push systems, which has also been confirmed by a large
number of the analyzed PCS selection studies (Sect. 2.2.3). And even if this would
not be true, the case of unlimited buffers can be emulated by setting a sufficiently
high WIP cap. For determining how much WIP is needed, the operating curve
concept of Nyhuis and Wiendahl (1999) provides a practically applicable approach,
which is conform with the Lean Manufacturing philosophy and is closely related to

Little's Law.[4] The physics and strategic considerations behind the second question can be found in Olhager (2003). The basic idea is to move the OPP as far upstream as permitted by the customer lead time. For the third question, usually Lean Manufacturing would recommend building a supermarket-based MTS system (Rother and Shook 1998), whereas MRP-based systems would resort to an MTF strategy (Orlicky 1975).

It is aimed at developing a PCS engineering framework that provides an integrated answer to all three questions. As it is obvious from the literature review, there is no one fits all solution and the choice of the PCS and its parameters is strongly influenced by its operating environment (Sect. 2.2.3). Therefore, using a generic PCS model, like the EKCS or the PAC system that can emulate different relevant PCS, seems to be the most promising approach. Hence, a queuing network based meta-model that provides, by optimization of its parameters, an answer to the three questions formulated above is developed. It should be described in a simple and intuitive way and the essence of the PCS that have proven successful in the past should be synthesized into it. In contrast to many existing PCS selection studies, it is tried to gain a holistic systems engineering perspective on the design drivers. Moreover, through explicitly puzzling out complexity reduction techniques, the approach should be able to handle the complexity of real life production systems. Emphasis is put on selecting the right objective function. Relevant operational metrics and trade-offs among them should be taken into account from a production management point of view. The objective function should thus be oriented at the analysis of the influences of PCS on operational performance as developed in Sect. 2.1.4.

The focus is then put on the third question. Here, a new perspective is provided by introducing a hybrid MTF/MTS approach, which is enabled by the proposed generic model. The approach will be combined with limited buffers between processes. It will be shown that under certain conditions, this approach is able to significantly outperform the pure strategies. Thereby, other than in existing studies that incorporate advance demand information (ADI), forecast error is explicitly addressed as part of the dynamic complexity in the PCS design.

From a methodological standpoint, the following work resorts to systems engineering (Sage and Rouse 1999), Operations Research (Neumann and Morlock 2002), decision analysis (Howard 1983), and simulation (Law and Kelton 2008). The relevant methods will be introduced when they are needed within the following chapters.

---

[4] For an introduction to Little's Law see Stocker and Waldmann (2003)

Chapter 3
A Queuing Network Based Framework for PCS Engineering

3.1 Design Drivers and Initial Complexity Reduction

3.1.1 Structuring Design Drivers

In order to develop a framework for PCS engineering, the relevant drivers that influence PCS design need to be elicited first. Two main driver categories can be identified: structure-based drivers and variability-based drivers. Structure-based drivers (1) stem from the static production setup, whereas variability-based drivers (2) stem from the dynamic behavior of the production system. Variability-based drivers can be further separated according to the source of the variability into drivers resulting from production system variability (2a) and drivers resulting from demand variability (2b). Drivers based on production system variability have their origin within the production system or its inputs. Drivers based on demand variability have their origin at the customer. Table 3.1 illustrates this split and gives concrete examples for each driver category.

A PCS engineering framework needs to be able to cope with all categories of drivers. With regard to PCS design, a helpful commonality of production system variability-based drivers is that all of them ultimately affect the cycle times of processes. Demand variability-based drivers ultimately effect the customer lead time and the streams of actually demanded and forecasted quantities. These connections will be exploited later when formulating a generic PCS model. Moreover, the sought after generic PCS model needs to be able to appropriately represent the structure-based drivers.

The design of a PCS naturally also includes position, type, and amount of inventory. The split of the drivers above into structure-, productions system variability-, and demand variability-based can also be applied to inventories. Stocks can be necessary to deal with structural circumstances (e.g. parts that are assembled into one product share a common process and therefore need to be stored) or can be necessary to act as a hedge against production system- or demand variability.

C. Karrer, *Engineering Production Control Strategies*, Management for Professionals, 37
DOI 10.1007/978-3-642-24142-0_3, © Springer-Verlag Berlin Heidelberg 2012

Table 3.1 Structure of design drivers

(1) Structure-based drivers		• Number of different (intermediate) products
		• Bill-of-material (BOM) structure
		• Product-process matrix
		• Line length
		• Variation of standard cycle times across processes (existence of a bottleneck)
		• Capacity and resulting utilization
		• Batch processes and optimal batch sizes
		• Shift synchronization
		• Point of variant determination
(2) Variability-based drivers	(2a) Production system variability	• Process breakdowns and minor stops
		• Process speed losses
		• Changeover times
		• Scrap and rework
		• Supplier delivery and quality performance
	(2b) Demand variability	• Order size and frequency
		• Customer lead time
		• Variability in demanded volume and mix
		• Reaction to unfulfilled orders
		• Forecasting error and frequency
		• Correlation among demands of products

3.1.2   Complexity Reduction by Defining Planning Segments

In order to engineer a PCS for a complex production system, it is helpful to reduce the complexity induced by structure-based drivers upfront to a level just not detrimental for the quality of the design. This can be achieved by the following two-step approach. First, products are grouped into families among which a detached PCS engineering is valid. Then, for each family, processes in the end-to-end process chain are aggregated into planning segments (PS). Figure 3.1 illustrates this arrangement of processes in PS.

The criterion for building product families is that production processes should not be shared among families. In other words, the family's boundaries have to be chosen such that equipment can be dedicated to them and no interference occurs. Within a family, one or more sequential processes are aggregated to a planning segment according to the following definition.

3.1.2.1   Definition of Planning Segment (PS)

A planning segment is a logical group consisting of (1) one or more processes that do not need separate planning or scheduling and thus are connected directly or only by standard in-process-stock (SIPS) and (2) an input buffer that establishes the link with preceding planning segments.

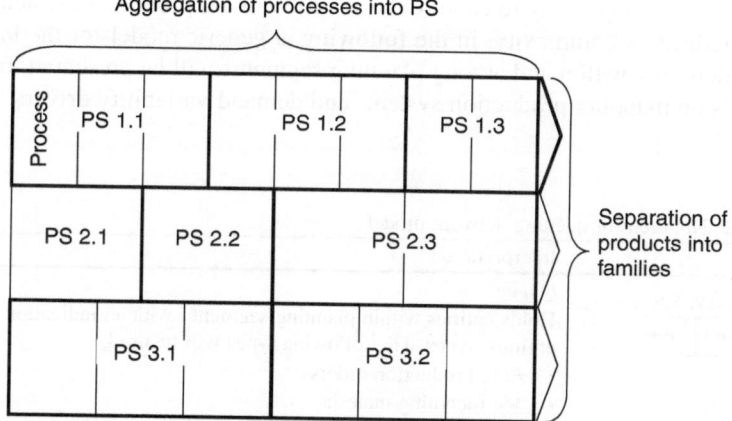

Fig. 3.1 Structuring the production system into planning segments (PS)

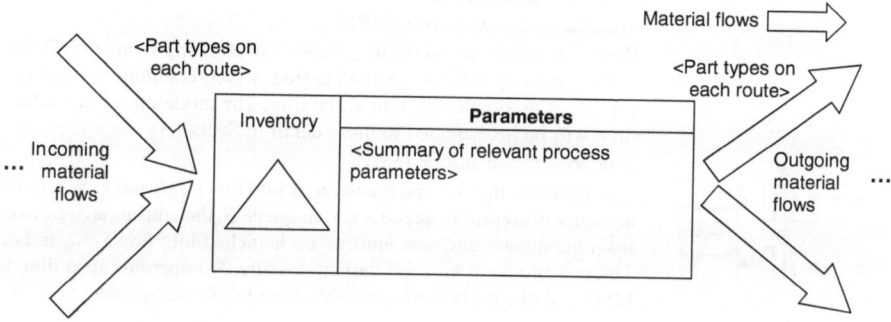

Fig. 3.2 Graphical representation for planning segments

In this definition, stock is usually considered as SIPS if the amount of WIP is only present to enable a better synchronization of the processes and has a reach of few cycles only. Exemplary reasons for which processes would need separate planning could be different shift models, inability to physically co-locate them, or a too large variation in cycle times driven either by production system variability or by different standard cycle times. In these cases, they obviously cannot be aggregated to one planning segment. The input buffer of a planning segment consolidates all buffers that feed into it and might physically be located separately. This consolidation further reduces structural complexity. To understand the structure of the production system under investigation, it is helpful to graphically map the identified planning segments and the flows among them. Figure 3.2 suggests a possible basic visualization for planning segments.

The resulting visualization can be used as communication tool in the engineering team and support the collection of relevant data.

After introducing a way to capture the structure of the production system and to upfront reduce its complexity, in the following, a generic model for the logic and information flow within and among planning segments will be presented, followed by details on mapping production system- and demand variability drivers.

Table 3.2 Notation for queuing network model

Symbol	Interpretation
<Type$_i$>	*Queue*[a] Holds entities within planning segment i with an indication of the entities' types. The following types will be used: • PO_i: Production orders • M_i: Incoming material • PC_i: Production clearance • IC_i: Internal clearance • IB_i: Internal buffer • FG: Finished goods
MP$_i$	*Manufacturing processes (MP)*[a] Represents the manufacturing processes in planning segment i. Delays entities passing it for a specified period, which is influenced by the drivers of production system variability. The modeling of the delay time will be investigated in more detail in Sect. 3.3
a → A → c B → d b → → e	*Synchronization station (SSt)*[a] Synchronizes flows a and b and feeds into flows c,d, and e. Whenever an entity is present in queue A and in queue B, the entities are removed from the queues and new entities are launched into flows c,d, and e. The synchronization is not variant-specific. *SSt* operates according to FIFO and always removes the oldest entity of each queue
a → A B variant specific → c b → → d → e	*Variant-specific synchronization station (VSSt)* Needed for the operation in multi-product environments. Works after the same principle as a synchronization station, however, two entities in queues A and B are only matched if they are of the same type. To find a match, each entity in queue A is compared with each entity in queue B. If several couples can be found, the FIFO logic is applied with a predefined queue priority. In the following application, queue PO_i is given priority. The induced entities in flows c,d, and e are of the type of the match
a → → ⊗ → b condition	*Conditional disposal (CD)* Based on a specified condition, disposes entities from flow a or routes them into flow b
Demand model Production orders F/Cs Orders	*Demand model* Generates flows of production orders that feed into queues PO_i. They can be either based on real customer orders or forecasts (F/Cs). At first, the demand model is considered as a black box, its detailed operation will be described in Sect. 3.4

[a]Adapted from Liberopoulos and Dallery (2000)

3.2 Generic Model Formulation

3.2.1 Notation

In order to describe and discuss a framework to engineer production control strategies, a common and formalized language needs to be defined. For the following queuing network formulation, the notation given in Table 3.2, which extends the notation of the unified framework for pull control mechanisms as proposed by Liberopoulos and Dallery (2000), is used.

3.2.2 Queuing Network Representation of Planning Segments

In the following, the generic model around which the proposed PCS engineering approach evolves will be described as queuing network model and then be analyzed in detail. The approach is titled 'generic' since it is capable, depending on the choice of its parameters, to emulate different PCS approaches. To explain the logic and basic properties of the suggested approach, the two last planning segments of a serial manufacturing process chain are displayed in Fig. 3.3. Its details and behavior will be described stepwise below.

For the following, it is assumed that the production system consists of planning segments $i \in \{1, ..., N\}$ and products $j \in \{1, ..., M\}$. The system's behavior is influenced by the initial contents of its queues. Table 3.3 lists the queues in a planning segment i together with their function and initial value.

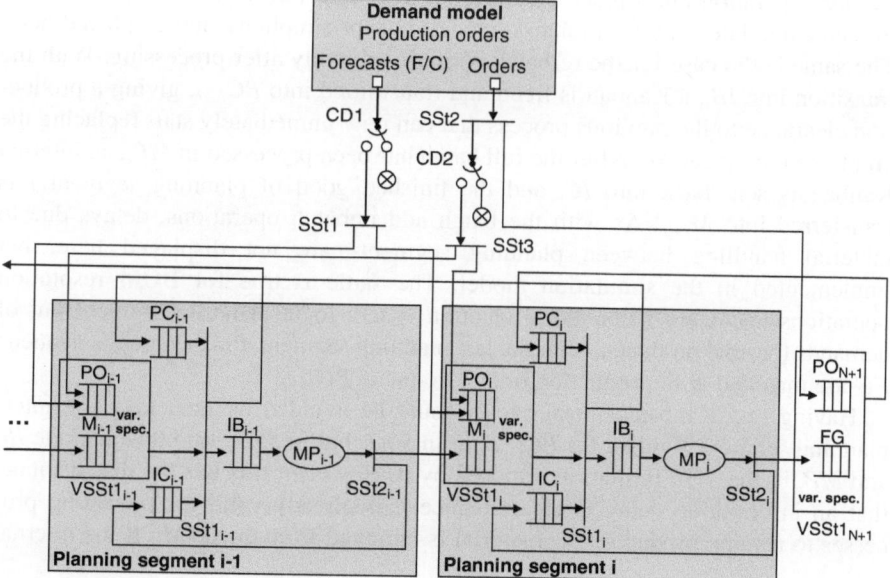

Fig. 3.3 Queuing network model of planning segments

Table 3.3 Queues and their initial values

Queue	Initial content	Purpose
PC_i	$K_i \in \mathbb{N}$	Stores free Kanbans representing the production clearance from the succeeding planning segment, the initial value Ki represents the initial number of free Kanbans in the system
IB_i	0	Technically necessary internal material buffer to realize unbatch operations and planning segments in which several units are processed simultaneously. Empty in initial system state
IC_i	$c_i \in \mathbb{N}$	Buffer for planning segment's internal production authorizations that define the capacity of IB_i. Initialized with the number of units processed simultaneously (c_i) in the planning segment
PO_i	0	Queue containing the production orders for the planning segment, initially empty
M_i	$\sum_{j=1}^{M} S_{ij} \in \mathbb{N}$	Input material queue, which is initialized with basestock S_{ij} of each product type j. For planning segments that process raw materials infinite material availability $\forall j : S_{ij} = \infty$ is assumed
FG	$\sum_{j=1}^{M} S_{N+1,j} \in \mathbb{N}$	Basestock of finished goods

At first, the logic implemented within one planning segment is explained. When a production order arrives in queue PO_i, $VSST1_i$ checks if the right type of material is available. Next, through synchronization station $SSt1_i$, it is checked if a Kanban is available in queue PC_i, which has been initialized with K_i free Kanbans, and whether the entity can be loaded into the process, signaled by a Kanban in queue IC_i. If this is the case, the entity is transferred into queue IB_i. Here, usually an unbatch operation takes place breaking up the batch into its single units. This is implemented later in the simulation model, but for simplicity, not displayed here. The same is the case for the re-batch operation directly after processing. With the transition into IB_i, a Kanban is freed and transmitted into PC_{i-1}, giving a production clearance to the previous process that can now immediately start replacing the empty spot in queue M_i. After the full batch has been processed in MP_i, an internal Kanban is sent back into IC_i and the finished good of planning segment i is transferred into M_{i+1}. As with the batch and unbatch operations, delays due to material handling between planning segments are not displayed here, but implemented in the simulation model. The same is true for BOM resolution operations necessary to create production orders for a planning segment out of demands for final products. After the last planning segment, finished goods in queue FG are matched with production orders in queue PO_{N+1}.

Having two Kanban control cycles could be avoided by launching the inter-planning segment Kanban for PC_{i-1} not in $SSt1_i$ but in $SSt2_i$, and thus include IB_i and MP_i in the WIP limitation imposed by it. However, this has the disadvantage that an information delay is created since it is already valid for preceding processes to resume production as material is removed from queue M_i. If the internal control loop would simply be omitted in the setting above, queues IB_i would have no de-facto WIP limit and thus, the overall WIP limitation would not be effective.

The production orders that feed into the planning segments are either based on forecasts, or based on actual orders. Forecast-based orders will be timed by the MRP logic in the demand model such that the following actual orders can be fulfilled just within the customer lead time to create neither delays nor unnecessary stock. Moreover, they are assumed to be perfectly leveled over the different variants. At this point, another parameter called 'forecast trust' $FCT_j \in [0, 1]$ is introduced. This parameter specifies to which extent the forecasted volumes are translated into actual production orders. As an example, for $FCT_j = 0.5$, half of the forecasted volume for variant j would be launched as production orders. In Fig. 3.3, this is indicated using a conditional disposal $(CD1)$, which would in the example of $FCT_j = 0.5$ dispose half of the volume ordered of type j. However, in the implementation of a simulation model, it is more convenient to generate production orders only for FCT_j percent of the forecasted volume in the first place.

The forecast-based production orders are then distributed by $SSt1$ to all planning segments located before the order penetration point (OPP). After the OPP, only actual customer orders are used. This corresponds to a pure make-to-order (MTO) system. Production orders based on actual customer orders are routed by $SSt2$ to PO_{N+1} to be matched with finished goods. Moreover, they are routed to $CD2$, where it is checked, if they were already covered by a forecast-based production order. Only if not, they are forwarded to the non-MTO planning segments upstream the OPP. This condition allows for an adjustment over time to over- or under-forecasting as it is performed by MRP. Thus, $CD2$ needs to evaluate, for each product j, the following condition (3.1).

$$\int_0^T PO_j^{F/C}(t)dt < \int_0^T PO_j^{O}(t)dt \tag{3.1}$$

T: Current point in model time
j: Product index
$PO_j^{F/C}(t)$:   Stream of production orders (rate) being launched based on forecasts for product type j at time t
$PO_j^{O}(t)$:   Stream of production orders (rate) being launched based on actual customer orders for product type j at time t

If the condition is true, entities will be forwarded, otherwise disposed.

The synchronization of customer orders and finished goods as it is modeled in $VSSt_{N+1}$ assumes that the customer will accept delayed orders, in other words, backlogging will be performed.

As introduced in Sect. 2.1.3, the OPP is an important design parameter in PCS engineering and will be incorporated as last parameter. In our model, OPP_j denotes for each product j, the planning segment in front of which production according to actual customer orders starts, thus $OPP_j \in \{1, ..., N + 1\}$. If OPP_j equals $N + 1$, it is located in front of the finished goods inventory. From the OPP downstream, an MTO system is indicated, which means that neither basestock S_{ij} in queues M_i (as introduced in Table 3.3), nor forecast-based production orders are needed. Therefore, it can be stated that

$$S_{ij} = 0, \quad \forall j = 1, ..., M, \; i > OPP_j \tag{3.2}$$

Moreover, planning segments after the OPP will not have their queue for production orders PO_i connected to $SSt1$ and hence do not receive forecast-based production orders. In the illustrating Fig. 3.3 above, the OPP is assumed to be at the end of the line ($OPP_j = N + 1$).

To summarize, the proposed queuing network model can, depending on the choice of its parameters K_i, S_{ij}, FCT_j, and OPP_j emulate several production control approaches. This includes hybrid strategies like for instance a hybrid make-to-forecast/make-to-stock strategy for controlling planning segments upstream the OPP. Such a hybrid system has not yet been explored, its advantages will be investigated later. Details on the solution space and the relation to other production control approaches will be given in Sect. 3.2.4. In order to better comprehend the coherences in the formulated queuing network approach, some basic properties will be revealed next.

3.2.3   Basic Properties

For the following closer examination of the properties of the proposed queuing network, function $I(.)$ is defined to model the current inventory held in queues or manufacturing processes. Moreover, it is assumed that the OPP is located in front of the finished goods inventory (3.3).

$$OPP_j = N + 1, \quad j = 1, ..., M \tag{3.3}$$

The following invariants explain the coherence between the WIP in the system and its parameters. They are proven by analyzing all possible system transitions by complete induction as it is commonly performed for queuing network models and also done by Dallery and Liberopoulos (2000). For the WIP control cycle located inside a planning segment, invariant 1 holds and the WIP is limited to c_i.

Invariant 1

$$I(IC_i) + I(IB_i) + I(MP_i) = c_i, \quad i = 1, ..., N \tag{3.4}$$

Proof: In the initial network state, no unit is being processed and thus $I(MP_i)$ = 0. Moreover, by definition (see Table 3.3), $I(IB_i) = 0$ in the initial state. From this, $I(IC_i) = c_i$ and (3.4) follows for the initial state. Looking at all possible transitions involving the queues and the manufacturing process of (3.4), it can be seen that whenever an entity is deducted from one summand of the left side of the equation, it has to be added to another one (3.5), and thus (3.4) holds.

$$
\begin{aligned}
&i) \quad I(IC_i) - 1 \Rightarrow I(IB_i) + 1 \quad i = 1, ..., N \\
&ii) \quad I(IB_i) - 1 \Rightarrow I(MP_i) + 1 \quad i = 1, ..., N \\
&iii) \quad I(MP_i) - 1 \Rightarrow I(IC_i) + 1 \quad i = 1, ..., N
\end{aligned}
\tag{3.5}
$$

A similar consideration can be made for the WIP between two connected planning segments.

Invariant 2

$$
I(M_{i+1}) + I(IB_i) + I(MP_i) + I(PC_i) = K_i + \sum_{j=1}^{M} S_{i+1,j}, \quad i = 1, ..., N
\tag{3.6}
$$

Proof: In the initial state, (3.6) follows directly from replacing the inventory values with the initial values from Table 3.3. Listing all possible transitions that affect the summands in (3.6) reveals that whenever an entity is deducted from one element of the left side of the equation, it is added to another one (3.7). Note that in invariant 2 and the following considerations, the initial material queue M_1 that contains the infinite raw material supply, is not included. Also the material queue of the planning segment after the OPP, here the FG queue, has to be excluded since its content is subject to forecast error.

$$
\begin{aligned}
&i) \quad I(M_{i+1}) - 1 \Rightarrow I(PC_i) + 1 \quad i = 1, ..., N - 1 \\
&ii) \quad I(PC_i) - 1 \Rightarrow I(IB_i) + 1 \quad i = 1, ..., N - 1 \\
&iii) \quad I(IB_i) - 1 \Rightarrow I(MP_i) + 1 \quad i = 1, ..., N - 1 \\
&iv) \quad I(MP_i) - 1 \Rightarrow I(M_{i+1}) + 1 \quad i = 1, ..., N - 1
\end{aligned}
\tag{3.7}
$$

With the help of invariant 2, we can derive an upper and a lower bound for the WIP within a production system[1] that follows the developed control approach. Rewriting (3.6) with all WIP containing queues on the left side leads to (3.8).

$$
I(M_{i+1}) + I(IB_i) + I(MP_i) = K_i + \sum_{j=1}^{M} S_{i+1,j} - I(PC_i), \quad i = 1, ..., N - 1
\tag{3.8}
$$

---

[1] Note that the presented WIP limits do not include raw materials and finished goods

Since parameters K_i and S_{ij} are constant during system operation, PC_i, the queue that contains the free Kanbans, is the only variable factor. The lower bound of $I(PC_i)$ equals 0, if no free Kanbans are present. The upper bound is, if all material from M_{i+1} has been consumed, $K_i + \sum_{j=1}^{M} S_{i+1,j}$. Thus, the upper bound WIP in the production system is given by

$$WIP_{\max} = \sum_{i=1}^{N-1} \left(\sum_{j=1}^{M} S_{i+1,j} + K_i \right) \tag{3.9}$$

and the lower bound is given by

$$WIP_{\min} = 0 \tag{3.10}$$

The considerations above are formulated under the assumption that the OPP is positioned after the last planning segment and thus, a finished goods inventory exists. Forecast-based orders for which no actual orders exist (a result of forecasting error) accumulate in the finished goods queue FG. If the OPP is positioned somewhere else within the process flow, the considerations above are still valid in a case where $FCT_j = 0$ $\forall j$. In any other case, goods produced based on forecasted orders for which an actual order is outstanding would accumulate in queue M_{OPPj} and remain as additional basestock until the balancing mechanism (3.1) clears them. In this case, the queue M_{OPPj} would have to be excluded from the invariants as already noted in the proof of invariant 2.

The chosen queuing network approach distinguishes between two types of inventory. Inventory determined by K_i and inventory determined by S_{ij}, which are both contained in queue M_i. This split conceptually corresponds to the decomposition of variability-based design drivers introduced in Sect. 3.1.1. Inventory driven by K_i covers for production system based variability. The basestock driven by S_{ij} covers for demand uncertainty, in other words, it enables satisfying customer orders with a shorter demanded lead time than the actual production lead time. It is worth noting that only inventory driven by the choice of K_i influences the production system lead time negatively by inducing waiting times. Besides, only one WIP initialization is variant-specific: S_{ij}. This split moreover conceptually corresponds to the first two different meanings of pull systems identified in Sect. 2.1.

3.2.4   Analysis of the Resulting Solution Space

In Chap. 2, three major questions that PCS engineering needs to answer were elaborated. The approach presented above addresses all of them. The question of the optimal WIP level between processes to cover for production system variability can be answered by optimizing K_i. The OPP is allocated by OPP_j. The OPP upstream control design is set up by optimizing FCT_j and S_{ij}.

Depending on the choice of FCT_j and S_{ij}, different control paradigms can be emulated as illustrated in Table 3.4.

Table 3.4 Emulated control paradigms

FCT_j	S_{ij}	Emulated paradigm
0	0	Make-to-order (MTO)
1	0	Make-to-forecast (MTF)
0	$\geq 0, \exists i,j : S_{ij} > 0$	Make-to-stock (MTS)
>0	$\geq 0, \exists i,j : S_{ij} > 0$	Hybrid MTF/MTS

Taking also K_i into account, popular existing PCS[2] can be emulated. Table 3.5 provides some popular examples.

Table 3.5 Examples for emulated PCS

FCT_j	S_{ij}	K_i	Emulated PCS
1	0	∞	Classic MRP, "Push"
0	0	$0 < K_i < \infty$	Kanban control system
0	≥ 0	∞	Basestock control system
0	0	$\infty, \quad i \in A \subseteq \{1,...,N\}$	Horizontally integrated hybrid PCS
		$0 < K_i < \infty, \quad i \in \{1,...,N\}\backslash A$	
1	0	$0 < K_i < \infty$	Synchro MRP

As mentioned, the presented approach is powerful in the sense that it is capable of emulating relevant PCS approaches. Thus, solutions can be engineered to the needs of the production system and its customers. However, the price for this is a large solution space. For planning segments in a serial flow and with unlimited raw material supply (thus K_1 is infinite), the number of parameters to optimize is described by expression (3.11).

$$
\underset{\substack{\Uparrow \\ K_i}}{(N-1)} \quad + \quad \underset{\substack{\Uparrow \\ FCT_j}}{M} \quad + \quad \underset{\substack{\Uparrow \\ OPP_j}}{M} \quad + \quad \underset{\substack{\Uparrow \\ S_{ij}}}{\sum_{j=1}^{M}(OPP_j - 1)} \tag{3.11}
$$

K_i needs to be defined between all N sequential planning segments. Thus, the number of parameters equals $N-1$. OPP_j and FCT_j need to be defined for all M products. The number of variables S_{ij} that need to be included depends, due to (3.2), on OPP_j. S_{ij} needs to be defined for all M products in front of each planning segment upstream the OPP, except for the first one, whose input buffer hosts the raw material supply, which is assumed to be infinite.

The resolution of variable OPP_j is $N + 1$ levels by definition. If the resolution with which parameters K_i, FCT_j, and S_{ij} should be determined is set to L levels, the size of the solution space equals (3.12).

$$
L^{(N-1)+M+\sum_{j=1}^{M}(OPP_j-1)} \cdot (N+1)^M \tag{3.12}
$$

---

[2] See literature survey in Sect. 2.2

For production systems with non-serial flows, $\sum_{j=1}^{M} (OPP_j - 1)$ in (3.11) and (3.12) needs to be replaced with the number of planning segments that are allocated upstream the OPP and do not process raw materials. If the number of intermediate products varies among planning segments, also this would have to be accounted for in the equations.

For an exemplary production system with 10 serial planning segments, 20 products, the OPP allocated at the end of the line, and a resolution of ten levels, (3.12) leads to a solution space with a size of roughly $10^{9+20+200} \cdot 11^{20} \approx 6.73 \cdot 10^{249}$. This example shows that already for very simple production systems with few products, the solution space reaches enormous magnitudes. Therefore, in the course of a PCS engineering process that will be developed in Chap. 4, a valid decomposition of the optimization problem will be derived and further approaches to avoid a combinatorial explosion in that magnitude will be suggested. This will make the PCS engineering problem solvable by complete enumeration.

In appendix 8.8, track will be kept of the respective minimal size of the solution space that is achieved with the proposed complexity reduction techniques. Next, the derivation of delay times for MPi in the presented queuing network approach, which are deducted from production system variability-based design drivers, will be explained. This is followed by the explanation of the demand model, which maps demand variability-based design drivers and has been treated as a black box so far.

3.3   Mapping Production System Variability

3.3.1   An Integrated Approach to Stochastic Modeling of Production System Variability

It can be concluded from previous research (e.g. Nyhuis and Wiendahl 1999) that production system variability is an important driver for PCS engineering. The following approach builds on the fact that all production system variability ultimately affects the cycle time of a planning segment and thus, in the proposed model, the delay time used in the element MP_i. The concept of the Overall Equipment Effectiveness (OEE)[3] will be used to structure and map production system variability-based design drivers into MP_i. To make the following illustration of the proposed concept easily accessible, the consideration of different variants will be left out.

---

[3] An introduction to the concept of the Overall Equipment Effectiveness (OEE) can be found in Muchiri and Pintelon (2008)

Definition "Standard Processing Time" (*SPT*)

The standard processing time (*SPT*) is the total time needed to complete one unit if no disturbances in the manufacturing process occur.

Definition "Standard Cycle Time" (*SCT*)

The standard cycle time (*SCT*) is the standard time between the completion of two consecutive units. If n units are processed simultaneously, each having the standard processing time *SPT*, then

$$SCT = \frac{SPT}{n} \tag{3.13}$$

SPT and *SCT* are both deterministic.

Definition "Actual Cycle Time" (*ACT*)

The actual cycle time (*ACT*) is a random variable representing the actually observed cycle time during the process. It consists out of the *SCT* and an additional random period representing losses in terms of the Overall Equipment Effectiveness (OEE) called '*OEEloss*'.

$$ACT = SCT + OEEloss, \quad \text{with } OEEloss \geq 0 \tag{3.14}$$

The *OEE* as performance measure can be calculated by (3.15) and reaches 1 (*OEEloss* = 0) if the process is fully utilized during the available time (*AvailableTime*) and all units (number of units produced *x*) are produced within their *SCT*.

$$OEE = \frac{SCT \cdot x}{AvailableTime} = \frac{SCT}{\frac{AvailableTime}{x}} \approx \frac{SCT}{E(ACT)} \tag{3.15}$$

If deviations from *SCT* occur, the OEE measure distinguishes three loss categories:

- *Availability losses*: breakdowns, maintenance, changeovers
- *Speed losses*: minor stoppages, reduced speed
- *Quality losses*: scrap and rework

The following diagram (Fig. 3.4) establishes, over the OEE structure, the link from production system variability-based design drivers to the *ACT*. With the help of this structure, later a (simulation) model can be set up.

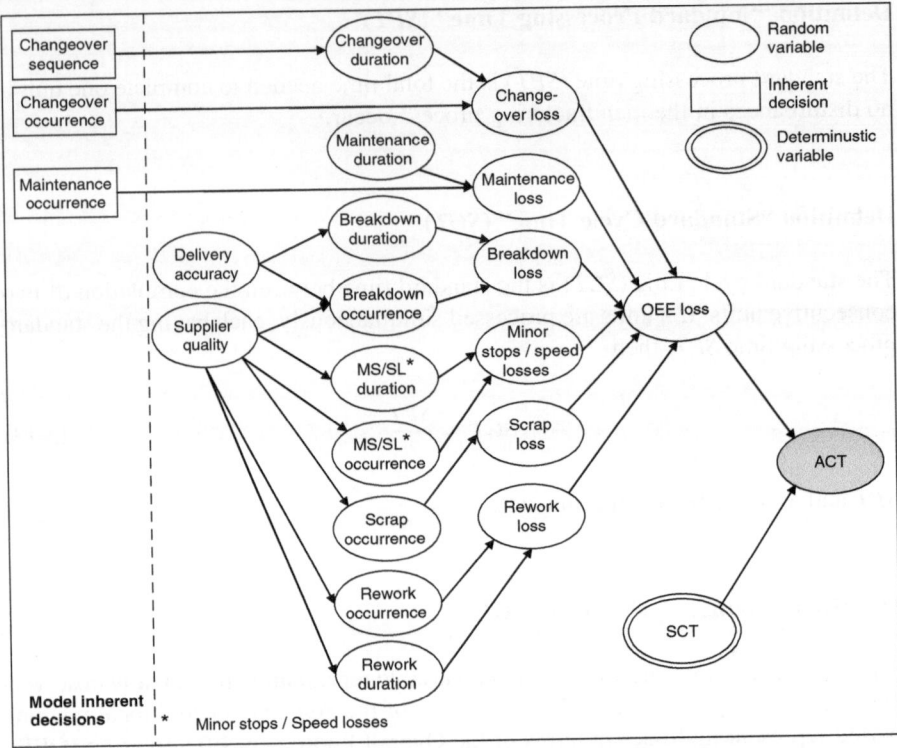

Fig. 3.4 Linking production system variability drivers to actual cycle times (ACT)

The total OEE loss depends on the losses in the different system-endogenous or exogenous (e.g. supplier caused) loss categories whose sum provides the *OEEloss* to be added to the *SCT* of a unit. The distributions of the random variables of each loss category depend on their occurrence and duration. The occurrence is either random or determined by certain system inherent conditions within the production system (e.g. next entity has a different type and thus causes a changeover loss). If the occurrence is a random variable, it can be coded either as the time span or as number of parts between two occurrences. The decision should be based on whether the occurrence is rather driven by time or by a certain amount of pieces produced. Moreover, the availability of data can have an influence on this choice from a practical point of view.

To determine the probability distribution of random variable *ACT*, either a direct distribution fitting on the OEE losses, or a separate modeling of each loss category's occurrence and duration probabilities are possible. However, the second approach should be the preferred one in most cases. Besides the usual lack of aggregated data in practice and the intricately shaped (e.g. multi-modal) distributions of *OEEloss*, it is not possible to examine the impact of changes in single production system variability drivers on the engineered PCS if the distribution is fitted directly.

We will not go into the details and challenges of fitting theoretical or empirical distributions for occurrence and duration probabilities here. The interested reader is referred to Law and Kelton (2008).

A helpful relation when fitting distributions in environments with a lack of data is the availability equation (Hopp and Spearman 2008).

$$Availability = \frac{E(Occurence) - E(Duration)}{E(Occurence)} \qquad (3.16)$$

Availability: Percent of the time the process is available
Occurrence: Random variable representing the time between two consecutive events that lead to non-availability (e.g. breakdowns)
Duration: Random variable representing the time duration of a non-availability event (e.g. breakdown)

Equation 3.16 is helpful for instance in cases, in which only data on duration and thus also the overall availability with respect to one loss category is tracked. Using the equation, a probability distribution for the occurrence with one parameter (e.g. the exponential distribution) can be hypothesized and fitted.

3.3.2 Definition of a Measure for Production System Variability

Next, a measure for production system variability is derived. This is necessary in order to be able to quantify the impact of production system variability on the production control strategy (PCS), and ultimately, to estimate the additional WIP and associated costs that are needed to cope with it. Using the previously introduced OEE measure (3.15) as indicator for production system variability has the downside that it only reflects the total loss and not its composition, which can be highly relevant. As an example, a 10% OEE loss could result either from many minor stops, or from one major stop. The effect on the production system of the latter would be worse since the induced disturbance is bigger, i.e. more planning segments and ultimately the customer are affected. Higher contingencies in the form of WIP are necessary to hedge against it. Therefore, a different approach will be taken and a loss function be defined for deviations from *SCT*.

A loss function assigns a loss value to each deviation from a specified target value. Common types of loss functions are binary-, linear-, or quadratic loss functions (Berger 1993). Figure 3.5 illustrates these three types.

A binary or "0–1"-loss function (a) would allow the parameter to vary in the interval *[T1, T3]*. If the parameter leaves this range, a constant loss is associated, no matter how far the parameter deviates. In the linear case (b), the loss grows linearly with the distance from *T2*. In the quadratic case (c), a quadratic loss growth is

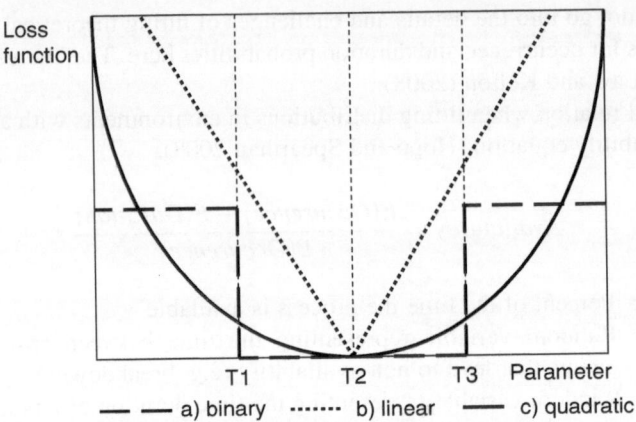

Fig. 3.5 Illustration of common loss functions

assumed. Quadratic loss functions are popular in quality control and referred to as "Taguchi loss functions" (Taguchi 1986).

To evaluate deviations from *SCT*, the construction of a quadratic loss function is appropriate since the longer the deviation from *SCT*, the more processes or planning segments are affected by the increasing negative consequences. Ultimately, even the customer can feel the effect as poor delivery performance. Equation 3.17 represents a quadratic loss function *L(X)* for random variable *X*, target value *T*, and a scaling constant *c* as proposed by Taguchi (1986).[4]

$$L(X) = c \cdot (X - T)^2 \qquad (3.17)$$

X: Random variable
T: Target value
c: Scaling constant

In the case of evaluating actual cycle times against their target cycle times, the loss function can be simplified to a function of the OEE loss by using (3.14).

$$L(ACT) = c \cdot (ACT - SCT)^2 = c \cdot OEEloss^2 \qquad (3.18)$$

The sought after variability measure VM_i of planning segment i is now computed as expected loss. Thereby the constant c is set to $\frac{1}{SCT}$ to generate a measure of relative variability, which enables the direct comparison of different planning

---

[4] Taguchi (1986) initially derived this loss function by approximating it with the Taylor series $L(X) = L(T + X - T) = L(T) + \frac{L'(T)}{1!}(X - T) + \frac{L''(T)}{2!}(X - T)^2 + ...$ Since $L(X)$ is minimal at $X = T$ and $L(T) = 0$ the quadratic term remains as most important one

segments or even production systems. If *SCT* differs among products in one
planning segment, the demand weighted average should be used. The measure
can be expressed by the first two moments of the probability distribution of the
OEE loss.[5]

$$VM_i = E(c \cdot OEEloss_i^2) = \frac{Var(OEEloss_i) + E(OEEloss_i)^2}{SCT_i} \qquad (3.19)$$

We define the variability measure of the whole production system *VMPS* with *N*
planning segments as aggregation of the variability of the single planning segments.

$$VMPS = \frac{1}{N} \sum_{i=1}^{N} VM_i \qquad (3.20)$$

VMPS expresses the variability or stability of a production system. *VMPS* can
also be considered as an indicator for how 'Lean' a production system is. Thus, it
could also be used as an indicator to benchmark different production systems within
the same industry against each other.

3.4 Mapping of Demand Variability

3.4.1 Stochastic Modeling of Demand Variability

In the PCS engineering framework introduced in Sect. 3.2, demand variability-based
design drivers are mapped using the 'demand model' element that generates streams
of forecast- and actual order-based production orders (POs). The parameters of the
demand model and their interrelations will be introduced with the help of the
illustrative Fig. 3.6, where cumulated POs are plotted for one product and one
forecasting period. Thereby, like in the demand model element presented earlier,
it is distinguished among production orders released based on forecasts and production orders released based on actual customer orders. For simplicity and better
readability, in the following explanations, the index *j* that would specify the respective product is left out.

The parameter time between forecasts (*TBFC*) stands for the period for which
the forecast value *FC* states the cumulated forecasted demand. If several forecast
values for a forecasting period exist, it is assumed that the most recent value
available at the beginning of the period is used. The parameter time between orders
(*TBO*) codifies the time between two consecutive customer orders. In Fig. 3.6 and

---

[5] Using $E(X^2) = Var(X) + (E(X))^2$ (explained for instance in Bosch 1998)

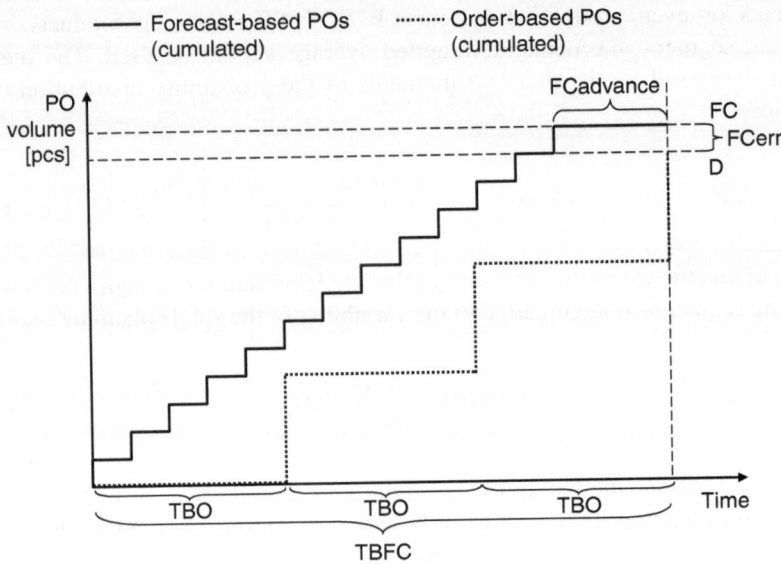

Fig. 3.6 Illustration of the production order streams over time

the following illustrations, *TBO* is constant, but could as well be a random variable. Each order has an associated customer lead time (*CLT*). It represents the time the customer is willing to wait for delivery after placing the order. Moreover, it is assumed that customer orders are multiples of batch size *BS*.

The demands during one period *TBFC* are cumulated into *D*. *D* is a random variable, since the demands occurring within the (next) forecasting period are unknown and variable. The periodic demands that occur are built-up by their corresponding production order stream $PO^O(t)$, which accumulates into the realization of *D* in one forecasting period, denominated d_t.

$$d_{c \cdot TBFC} = \int_{c \cdot TBFC}^{(c+1) \cdot TBFC} PO^O(t) dt \quad \forall c \in \mathbb{N} \tag{3.21}$$

$PO^O(t)$:   Stream of production orders (rate) being launched based on customer orders over time *t*

d_t: Cumulated demand in forecasting period beginning at *t*

The orders occur at discrete points in time within the forecasting period. Let *OP* be the set of discrete order points $OP = \{op_1, \ldots, op_R\}$ in one forecasting period. It is assumed that even though *TBO* can vary (if it is not deterministic), the demand rate within one forecasting period is constant.

$$\int\limits_{op_k}^{op_{k+1}} PO^o(t)dt = TBO_{op_k}^{op_{k+1}} \cdot c \quad \forall k = 1, ..., R-1 \tag{3.22}$$

$PO^o(t)$: Stream of production orders (rate) being launched based on customer orders over time t

op_k: Order points in one forecasting period

R: Number of order points in one forecasting period

c: Constant representing the demand rate in forecasting period

$TBO_{op_k}^{op_{k+1}}$: Time between orders of order points op_k and op_{k+1}

This assumption simplifies the following formulation and implementation of a demand model that creates corresponding streams of customer order- and forecast-based production orders. It enables the emulation of an MRP logic by evenly distributing the forecast-based production orders over $TBFC$, as also sketched out earlier in Fig. 3.6. The assumption is valid in practice and imposes no limit on generality but simplifies implementation.

In this context, $FCadvance$ determines how much in advance production orders based on forecast are released in order to be able to satisfy customer orders at the OPP. The forecast error ($FCerr$ in Fig. 3.6) is rooted in the difference between demanded and forecasted volumes $D - FC$.

After introducing the basic parameters, their relations and the basic assumptions made in the demand model, next, it will be explained how streams of forecast- or customer order-based production orders are generated in the demand model to enable simulation. Parameters CLT, BS, and $TBFC$ are assumed to be deterministic. TBO can be either deterministic or a random variable. D, FC, and $FCerr$ are random variables. The challenge lies in creating stochastic PO-streams that fulfill the interrelations of parameters described above, especially with respect to forecast error.

The chosen approach starts with random variable D for which, based on historical data, a theoretical or empirical probability distribution has to be fitted. If a value d_t is sampled from D, it represents the aggregated demand in the forecasting period that starts at time t.[6] From this, the demand rate c, which is assumed constant within the forecasting period (3.22), can be calculated as quotient $c = \frac{d_t}{TBFC}$. To generate orders, TBO is sampled (tbo) and the ordered quantity ($OrderSize$) is calculated with the help of term (3.23). The result is aligned with the assumption that its value is a multiple of BS.

$$OrderSize = \left\lceil \frac{tbo \cdot \frac{d_t}{TBFC}}{BS} \right\rceil \cdot BS \tag{3.23}$$

---

[6] Variables in lower case that are named after random variables represent, depending on the context, either realizations of the random variable, or observed values from reality used to fit a distribution for the random variable.

Next, the forecast error present in practice needs to be quantified for the model. Therefore, a probability distribution is fitted for random variable $FCerr$ from observed realizations $fcerr_t$. Two approaches are possible. The forecast error can be modeled either as absolute or relative deviation, depending on whether the size of the demand influences the magnitude of the forecasting error (Alicke 2005). In most cases, a relative modeling should be appropriate, which is assumed for the formulation of the equations in the following. Thus, if d_t and fc_t are observed values

$$fcerr_t^{relative} = \frac{d_t - fc_t}{d_t} \text{ (and } fcerr_t^{absolut} = d_t - fc_t) \qquad (3.24)$$

It is assumed that the forecast error is unbiased meaning that

$$E(FCerr) = 0 \qquad (3.25)$$

A potential forecast bias can easily be removed by deducting its expected value from the observations. In the relative case with T observations, this would lead to

$$fcerr_t^{relNoBias} = \frac{d_t - fc_t}{d_t} - \frac{1}{T} \sum_{i=1}^{T} \frac{d_i - fc_i}{d_i} \qquad (3.26)$$

Based on the resulting values from (3.26), a distribution fitting for $FCerr$ can be performed. In the absolute case, the bias is removed accordingly.

For random variable FC, a distribution fitting is not performed. The realizations in the simulation are calculated from the realizations of random variables D and $FCerr$. At this point in time, parameter FCT from the queuing network model can be incorporated and the result be rounded to full batches

$$fc_t = \left[\frac{(d_t + d_t \cdot fcerr_t) \cdot FCT}{BS} \right] \cdot BS \qquad (3.27)$$

The resulting production orders of batch size BS are pre-drawn by $FCadvance$ and evenly distributed over the forecasting period. Due to the constant demand rate assumption (3.22), this is conforming to the MRP logic. The considerations above are made for each product separately. However, if several products exist, the sequencing of the variants needs to be considered. Then, a 'leveled' system is used. This means that the orders are diversified as much as possible. As an example, for a three product setup (A,B,C) with equal demands for each variant, the demand model generates a stream of forecast-based POs '$A,B,C,A,B,C,\ldots$'. This assumption is recommended by the Lean philosophy (Womack et al. 2007) for many discrete batch production setups. However, there certainly are also cases where a more sophisticated sequencing could create value. But as we will see in Chap. 5, even in these settings, the validity of the statements that will be made for FCT and S_{ij} is not endangered.

Parameter *FCadvance* will be determined numerically in the course of the optimization procedure presented in Chap. 4.

3.4.2   Excursion: Possible Involvement of Customers

In the following, the presented demand model is viewed within a broader context and options how to involve the customer in the PCS engineering effort are discussed. In the approach described above, *FCerr* is fitted according to values observed in the past. This assumption is then encoded in the demand model and used to engineer a PCS. Here, a different approach could be taken that proactively manages deviations from forecasted demands. Instead of using spot estimates as forecasted demands like it is done in traditional forecasting, production and the customers (or at least the sales department) could agree upon a certain 'flexibility range' in form of a probability distribution of allowed deviations from the forecast. These flexibility ranges could then be used as input to the demand model and replace *FCerr*. This would ensure that the PCS is engineered in line with the customer's expectations and not based on a historic error measure. As long as the customer stays within his flexibility range, delivery is ensured by the design of the PCS. If he leaves it, the contingencies built into the PCS cannot ensure delivery anymore.

Based on this explicit agreement of forecasting ranges with the customer, a demand variability-based pricing could be realized. This would enable a diversified pricing as it has also been mentioned by Simchi-Levi et al. (2007). However, this idea will not be covered in more depth here and is left to further research.

Having specified a queuing network based framework for PCS engineering that is able to accommodate the identified design drivers, the next step is the development of an approach to optimize its control parameters that form the solution space.

Chapter 4
Numerical Optimization of Control Parameters Along a PCS Engineering Process

4.1 Objective Function Design

4.1.1 Selection of Valuation Measures

In order to optimize the parameters of the proposed queuing network model, as in any optimization problem, an objective function needs to be defined. The challenge lies in choosing the right metrics and in assessing the trade-offs among them. In the analysis of PCS selection literature (Sect. 2.2.3), no best practice approach could be identified. Hence, the selection of metrics is built on the analysis of the influence the PCS has on operational performance, which has been presented in Sect. 2.1.4.

According to Fig. 4.1, the PCS selection has a direct influence on WIP (by determining location, type, and amount) and delivery performance (e.g. location of order penetration point). The amount of WIP is the driver for several other directly or indirectly related operational metrics (lead time, quality, productivity, and so on). A helpful fact is that except for productivity, all of them become worse with increasing WIP levels and can therefore be captured by measuring WIP. Since this is not true for the influence of WIP on productivity, there is a trade-off that needs to be taken into account. This suggests that it is appropriate to use the WIP, productivity, and the delivery performance as metrics to create a comprehensive measure for the utility of a PCS configuration.

Looking at WIP, in the previous chapter it was noted that two types can be distinguished. WIP driven by K_i to cover for production system variability, and WIP driven by S_{ij} to cover for demand variability. Their influence on operational performance has to be evaluated by two differently parameterized value functions since some of the negative consequences of WIP, like increased lead times, occur only for K_i driven WIP.

K_i driven WIP acts as buffer and influences productivity. This effect has been explored in detail by Nyhuis and Wiendahl (1999) under the name 'operating curve'. To be able to value the WIP/productivity trade-off correctly, the relation of demand and available production capacity needs to be taken into account.

C. Karrer, *Engineering Production Control Strategies*, Management for Professionals, DOI 10.1007/978-3-642-24142-0_4, © Springer-Verlag Berlin Heidelberg 2012

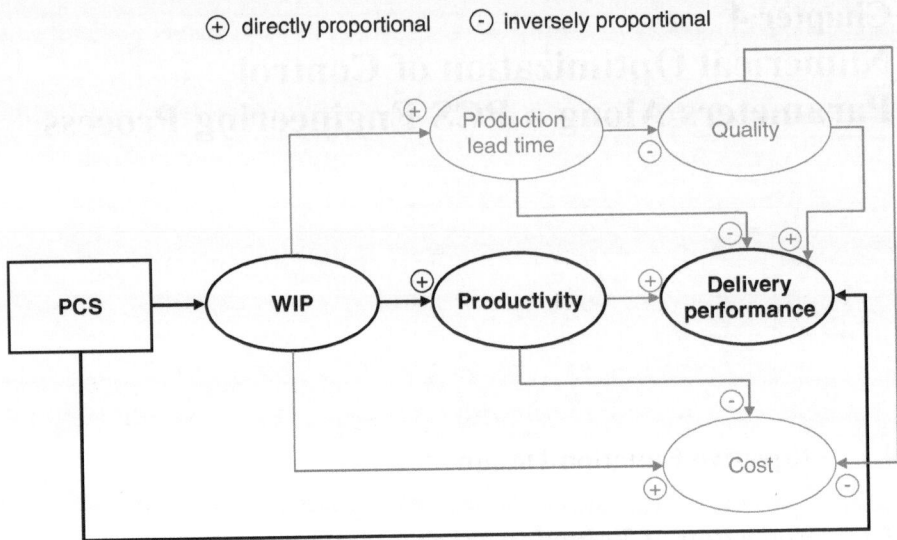

Fig. 4.1 Selection of valuation measures

Only in a scenario in which demand exceeds the available capacity, more WIP could generate more turnover and thus create additional value. In this situation, it is necessary to trade-off additional WIP against additional turnover. In a scenario in which capacity exceeds current demand, the situation is different. First, each company would try to reduce its capacity to match demand (starting with its most expensive capacity drivers). However, in many cases, this cannot be easily done; at least not in the short term (e.g. machines installed, labor laws). In this case, the value of productivity is not increased by more WIP since the time that would be gained by less congestion or missing material stops is consumed by idle times.[1] Therefore, an ideal PCS would reduce WIP to the level, at which the productivity is barely sufficient to satisfy demand and benefit from lower WIP levels. Details on this optimization follow in the description of the PCS engineering process (Sect. 4.3).

4.1.2 Value Analysis

Next, a value function for the three measures WIP, delivery performance, and productivity has to be defined. Due to its relative simplicity and good practical applicability, an additive value function as recommended for instance in Sage and Armstrong (2000) for problems in systems engineering will be considered.

---

[1] or in the case of manual labor, usually by speed losses

The general form of an additive value function $V(x_1,...,x_N)$ of the realizations x_i with single dimensional value functions $v_i(x_i)$ and weights w_i is given in (4.1). Thereby, N stands for the number of included dimensions.

$$V(x_1,...,x_N) = \sum_{i=1}^{N} w_i v_i(x_i), \quad v_i(x_i), w_i \in [0,1] \tag{4.1}$$

The single dimensional value functions $v_i(x_i)$ and the weights w_i are normalized to the interval [0,1] and w_i are chosen such that $\sum_{i=1}^{N} w_i = 1$ holds.

The application of an additive value function is subject to certain assumptions. In general, the additive value function is based on cardinal utility (Keeney and Raiffa 1993). This does not constitute a limitation in the considered environment since also all evaluated measures (WIP, delivery performance, and productivity) can be interpreted as cardinal type.

Sage and Armstrong (2000) mention mutual preference independence as an important assumption. This basically means that the preference for a specific outcome in one dimension is not influenced by the outcome of any other dimension. Due to the split of the optimization problem that will be presented in Sect. 4.3, the couples WIP/productivity and WIP/delivery performance need to fulfill mutual preference independence. As described in the discussion in Sect. 4.1.1, the WIP dimension measures the negative consequences or 'cost' of WIP (e.g. handling effort or capital expenditure). The negative value associated with WIP is independent of the achieved delivery performance and productivity, even though a causal coherence exists. Also the values ascribed to productivity and delivery performance can be assumed to be independent from the value of WIP. In contrast, for the relation of productivity and delivery performance, which has not to be considered in the optimization procedure, mutual preference independence does not necessarily hold. In the case in which customers do not accept delayed shipments, the value attributed to productivity is influenced by the achieved delivery performance.

In addition to mutual preference independence, Keeney and Raiffa (1993) mention additive independence as an assumption of additive value functions. This implies, within the considered ranges of the attributes, full mutual exchangeability of the objectives. The condition can be tested by using a hypothetical lottery as proposed in Keeney and Raiffa (1993). However, in practical applications this approach can be hard to convey. An alternative is provided by Delquie and Luo (1997). They formulate a simple trade-off condition that needs to be fulfilled from the production management's point of view in order to postulate additive independence. In the case in which additive independence cannot be assumed, another summand representing the interaction effect needs to be added into the value function (given a value function allowing for two dimensions). However, this should be seldom the case as long as the attributes are considered within reasonable ranges.

For further discussions of the assumptions and more details on the so-called multi attribute utility theory (MAUT) approach, which includes possible proceedings to determine the weights w_i, the reader is referred to relevant literature like Bamberg et al. (2008), Sage and Rouse (1999), or Keeney and Raiffa (1993).

Note that within this PCS engineering framework, if the specific case requires it, also any other objective function can be applied. However, the practical value of complex objective functions should be evaluated carefully.

Next, suggestions for the single dimensional value functions for delivery performance, WIP, and productivity will be presented.

4.1.2.1 Value Function Design for Delivery Performance

Delivery performance is commonly expressed either as percentage of orders that are delivered on time and in full ('alpha service level'), or as percentage of pieces that are delivered on time ('beta service level' or 'fill rate') (Alicke 2005). The due date on which the measures are based on is in practice mostly the first confirmed date, however, different dates can be used. The decision for a measure should mainly depend on the customers' needs and their way to measure delivery performance. When discussing the utility of a PCS in terms of delivery performance, it is reasonable to assume that in most cases, utility does not increase linear with delivery performance beginning from 0, but that a threshold exists, below which the PCS can be considered no value at all. The same way, since many companies agree a certain delivery performance target with their suppliers, also an upper threshold exists. Exceeding this threshold would not add significantly more value. Between the two thresholds, it is reasonable to assume that the utility increases linear. Based on these considerations, a generic, piecewise linear utility function with two threshold parameters a and b as depicted in Fig. 4.2 is suggested.

The two threshold parameters a and b depend mainly on how the market in which the considered company operates is organized. Monopoly or oligopoly markets can be expected to have a flatter curve than commodity markets. Moreover, agreements with the customers regarding delivery performance can play an important role. In PCS engineering, a and b have to be defined by the relevant

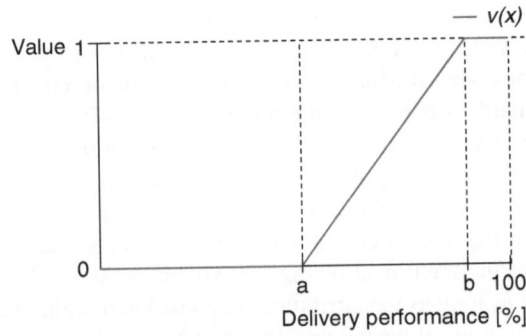

Fig. 4.2 Stepwise linear value function for delivery performance

stakeholders of the company. The value function can be expressed by the following equation.

$$v(x) = \max\left[0, \min\left(1, \frac{1}{b-a}x + \frac{a}{a-b}\right)\right] \text{ with } a, b \in [0, 1], \quad a < b \quad (4.2)$$

a: Minimum threshold below which PCS has no value in terms of delivery performance
b: Maximum threshold; exceeding this level will not increase the perceived value by the customer

4.1.2.2   Value Function Design for WIP

As argued before, WIP has positive and negative implications. On the one hand, increasing WIP can have a positive causal influence on productivity and delivery performance. On the other hand, WIP is not for free and has a certain cost associated with it. To be able to quantify the resulting trade-offs, here, the 'cost' of WIP will be brought into the equation. WIP is commonly measured either in pieces or in a monetary value. For PCS engineering, which is technical in nature, measuring in pieces is more beneficial. Since from the 'cost' point of view, increasing WIP leads to decreasing utility, a decreasing value function is assumed. Moreover, it is assumed that the function is linear and hence definable by a minimum WIP level *a* and a maximum WIP level *b*. The linearity assumption for the WIP's 'cost' is reasonable within certain bounds, as long as for instance no extra space needs to be rented and the handling effort does not explode due to the WIP being an obstacle for the material flow. Anyhow, if the linearity assumption seems not justifiable for the specific case, the value function for the 'cost' of WIP can also to be formulated more complex in order to better represent reality. Figure 4.3 illustrates the proposed linear function.

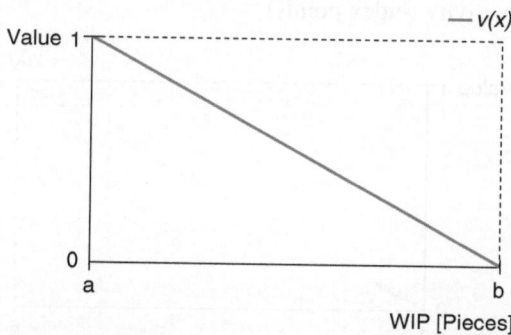

Fig. 4.3 Linear value function for WIP

The mathematical representation of such a function is given in (4.3).

$$v(x) = 1 - \frac{x - a}{b - a} \qquad (4.3)$$

a: Minimum WIP
b: Maximum WIP

The maximum WIP level *b* can be determined by space constraints and plausibility considerations in the simulation model. The minimum WIP level *a* can also be determined from the simulation model by varying the parameters under investigation and observing the achievable minimum.

4.1.2.3 Value Function Design for Productivity

If the plant operates in a situation in which demand exceeds the available capacity, it can be beneficial to accept an increase in WIP for an increase in productivity. Therefore, the increase in productivity needs to be valued. Higher productivity leads to higher sales. Productivity is commonly calculated as the ratio of what has actually been produced divided by the theoretical maximum production volume (with no variability in the processes). A mathematical definition of productivity will be presented later in Sect. 4.3.2. Let *a* denominate the minimum productivity (in productivity index points) and *b* the maximum productivity. *a* and *b* can be read from the operating curve of the system. The derivation of the operating curve requires numerical methods and will be presented in Sect. 4.3. Assuming a linear value increase (each unit that is sold extra has the same contribution margin), the value function would look like illustrated in Fig. 4.4.

The corresponding mathematical representation of the value function is shown in (4.4).

$$v(x) = \frac{x - a}{b - a} \qquad (4.4)$$

a: Minimum productivity (index points)
b: Maximum productivity (index points)

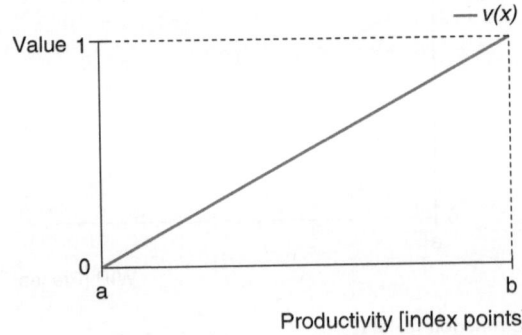

Fig. 4.4 Linear value function for productivity

As stated in the argumentation before, in the case in which capacity exceeds demand, the value of additional productivity is zero.

After defining value functions for all three metrics, the determination of the weights w_i remains in order to complete the necessary multi attribute utility function. As already mentioned, due to the decomposition of the optimization problem that will be proposed later in Sect. 4.3, it is not necessary to resort to the full utility function including all three attributes, but only to parts of it separately. The weights that need to be derived for the sub-functions will be named during the PCS engineering process. To determine the weights in these settings, it is recommended to follow a procedure common in decision analysis and described for instance in Sage and Rouse (1999).

4.2 Implementation of a Supporting Simulation Framework

4.2.1 Introduction to Discrete-Event Simulation and the Platform 'AnyLogic'

In Chap. 3, we defined an engineering framework to build production system models, which shall be optimized according to the objectives discussed in the previous section. Systems can be studied and optimized either analytically or numerically. An analytical study leads to an exact solution but can only be applied to models that are simple enough. If analytical study is not possible and systems are studied numerically, a computer is used to simulate them and to approximate the solution. Numerical studies have the advantage to be also applicable to complex systems (Law and Kelton 2008). Due to their complex structure and stochastic nature, the systems that are built using the proposed PCS engineering framework will be solved numerically. Discrete-event simulation is chosen to study them numerically. Discrete-event simulation maps a "system as it evolves over time by a representation in which the state variables change instantaneously at separate points in time" (Law and Kelton 2008). It was one of the goals of this research to also implement a corresponding simulation framework to support PCS engineering and to make the framework easily applicable for practitioners. The simulation framework enables the PCS engineer to map the system by using predefined objects (planning segments and the demand model) that are parameterized over a graphical user interface.

As simulation platform, AnyLogic from XJ Technologies (see Appendix 8.13) is used. AnyLogic is a Java[TM] and Eclipse[2] based modeling framework. It integrates discrete-event, agent-based, and system dynamics simulation and is therefore called

---

[2] http://www.eclipse.org/

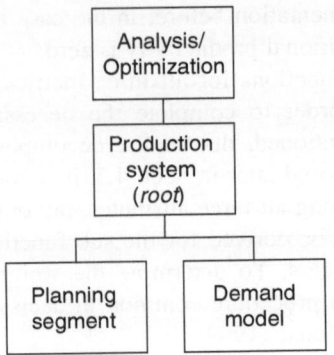

Fig. 4.5 Object hierarchy of the simulation model

a 'multi-paradigm' simulation framework. It incorporates the OptQuest$^{\text{TM3}}$ solver and enables fully object-oriented model building (AnyLogic Manual 2009).[4] The object-oriented approach enables the development of reusable and customizable components, such as planning segments. The built models can be converted into Java applets and thus be deployed as web applications. The modeling in AnyLogic is done on a graphical user interface on the one hand, and by directly entering Java code on the other hand. Basic modeling elements for the mapping of queuing networks, such as queues and delays, are predefined but can also be customized by changing their source code.

The subsequently mentioned guidelines were considered for the implementation of the simulation framework for PCS engineering. The developed objects (i.e. planning segments, demand model) can be fully parameterized and reused without any programming effort, which makes the framework easy to customize to any production setting, even for users without deep programming knowledge. For the developed objects, a convenient graphical representation has been developed, which helps the user of the model to observe its state and eases model validation. The whole implementation is strongly focused on computing efficiency to be able to realize fast and automated replications during optimization and sensitivity analysis. This is mainly achieved by using simple data- and control structures and by reducing data access to a minimum. The graphical representation can be switched off. Also the computation of metrics that are not needed for specific analyses can be disabled, which again increases runtime efficiency.

The object oriented simulation framework is developed in a hierarchy with three layers as illustrated in Fig. 4.5.

The bottom layer contains objects that represent planning segments and the demand model. These objects do not require any structural changes if they are used to build different production systems. Their behavior is completely controlled

---

[3] http://www.opttek.com/

[4] http://www.xjtek.com/

by parameters. Instances of both objects are used within the production system model, also called the 'root' object in AnyLogic. Here the structure of the production system is mapped by wiring planning segments and the demand model. Moreover, the root object is responsible to record performance metrics and for calculating utility functions. Above the root object, classes that execute the simulation once or several times in row in order to perform analyses or optimizations are allocated.

In the following, the three major objects: planning segment, demand model, and root will be outlined.

4.2.2   Implementation of Planning Segments

The icon with which planning segments appear in the production system model is displayed in Fig. 4.6.

According to the developed queuing network model, the planning segment implementation has ingoing ports for production orders, materials, and Kanbans. Outgoing ports exist for material and Kanbans. The icon also lists important indicators like the current WIP level in the planning segment, its OEE, the variability measure (as defined in Sect. 3.3.2), and the scrap rate.

The parameters that are used to customize a planning segment are summarized in Table 4.1. Thereby, a feature not existing in standard Java and worth noting here is that it is distinguished among static and dynamic parameter resolution. Parameters that are resolved in static mode are evaluated once at their first call and keep this value until the end of the model runtime. Parameters that are resolved in dynamic mode are reevaluated at every call. This is relevant if for instance a probability distribution is assigned to a parameter, which is frequently the case in the following. In dynamic resolution mode, each parameter call represents a sampling of the distribution. In static resolution, the distribution is only sampled at the first call and the resulting value is kept.

Subsequently, indexes j again represent different product types processed by the planning segment.

Icon

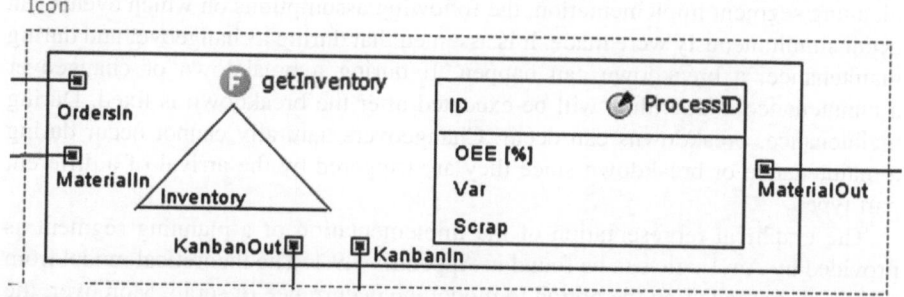

Fig. 4.6 Planning segment icon

Table 4.1 Parameters of the planning segment object

Parameter	Resolution	Data type[a]	Description
processID	Static	Integer	Identifier for planning segment
changeoverTTRdist	Dynamic	Double	Changeover duration distribution
maintenanceTTRdist	Dynamic	Double	Maintenance time to repair distribution
maintenanceTBFdist	Dynamic	Double	Maintenance time between failures distribution
scrapPBFdist	Dynamic	Integer	Scrap parts between failure distribution
reworkPBFdist	Dynamic	Integer	Rework parts between failure distribution
reworkTTRdist	Dynamic	Double	Rework time to repair distribution
minorStopPBFdist	Dynamic	Integer	Minor Stop parts between failure distribution
minorStopTTRdist	Dynamic	Double	Minor stop time to repair distribution
breakdownTBFdist	Dynamic	Double	Breakdown time between failure distribution
breakdownTTRdist	Dynamic	Double	Breakdown time to repair distribution
transportBatchsize	Static	Integer[j]	Batch size after processing for each product j
SCT	Static	Double[j]	Standard cycle time for each product j
setupFamily	Static	Integer[j]	Assignment of a setup family to each product j
shiftUptime	Static	Double	Continuous uptime period of planning segment before downtime
shiftDowntime	Static	Double	Continuous downtime period of planning segment before uptime

[a]Standard Java data types used in the following; [a,. . .,z] represents an array with dimensions a,. . .,z

An assumption that should be realistic for most applications has been made on whether the occurrence of OEE loss events is codified by parts or time between occurrences. In case the assumption does not hold, it can easily be changed in the source code of the planning segment object's class. A challenge is the consideration of dependencies among the occurrence of OEE loss events. Events that are measured using parts between failure (PBF) like scrap, rework, and minor stops do not interfere with events that are measured with the time between failure (TBF) approach like breakdowns or maintenance. This is the case since during a TBF driven event, also the count of produced parts stops. The events using the PBF approach can also be treated independent among each other since theoretically, a minor stop, rework, and scrap events can occur for one single part. For the TBF driven events, it is required that the TBF data used to fit the respective distribution is adjusted for other TBF driven events like for instance maintenance. In the planning segment implementation, the following assumptions on which events can occur simultaneously were made. It is assumed that during a changeover and during maintenance, a breakdown can happen. If during a breakdown or changeover a maintenance is due, they will be executed after the breakdown is fixed. During maintenance, breakdowns can occur. Changeovers naturally cannot occur during a maintenance or breakdown since they are triggered by the arrival of a different part type.

The graphical representation of the implementation of a planning segment as provided by AnyLogic can be found in Appendix 8.9.1. The theoretical model from Chap. 3 is extended by the option to model the occurrence of scrap. Moreover, the batch and unbatch operations that were not displayed in the formal queuing network model are included. To validate the parameterization of the planning segment, both, a plot of the OEE over time, and a plot of the variability measure over time are provided.

4.2.3   Implementation of the Demand Model

The icon representing the demand model on root level is displayed in Fig. 4.7.

The demand model has two outgoing ports, one for production orders based on forecasts, and one for production orders based on customer orders. Moreover, for each port, the number of entities that already left it is tracked and visualized. The demand model is customized using the parameters listed in Table 4.2.

To simplify the implementation of the demand model, it is assumed that the demand rate is not only constant over a forecasting period, but also within the time between two consecutive orders. This implies that at the point in time at which the demand rate changes, which is at the end of a forecasting period, also a customer order is placed. Thus, it is avoided that the model needs to account for different

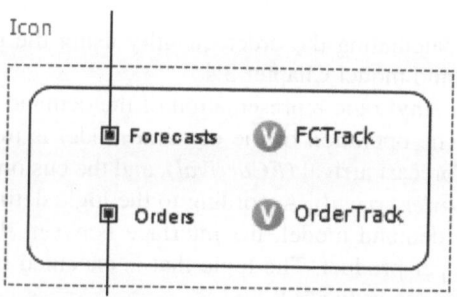

Fig. 4.7 Demand model icon

Table 4.2 Parameters of the demand model object

Parameter	Resolution	Data type[a]	Description
batchsizesOutbound	Static	Integer[j]	Batch sizes for each product j of which customer orders are multiples of
customerLeadtime	Static	Double[j]	Customer lead time for each product j
TBO	Dynamic	Double[j]	Distribution or value of time between orders for each product j
FCerror	Dynamic	Double[j]	Forecast error distributions for each product j
TBFC	Static	Double	Time between forecasts
demandInput	Dynamic	Double[j]	Demand distributions for each product j
FCadvance	Static	Double	Forecast advance period
productCount	Static	Integer	Number of products j
FCtrust	Static	Double[j]	Forecast trust parameter for each product j

[a]Standard Java data types used in the following; [a,. . .,z] represents an array with dimensions a,. . .,z

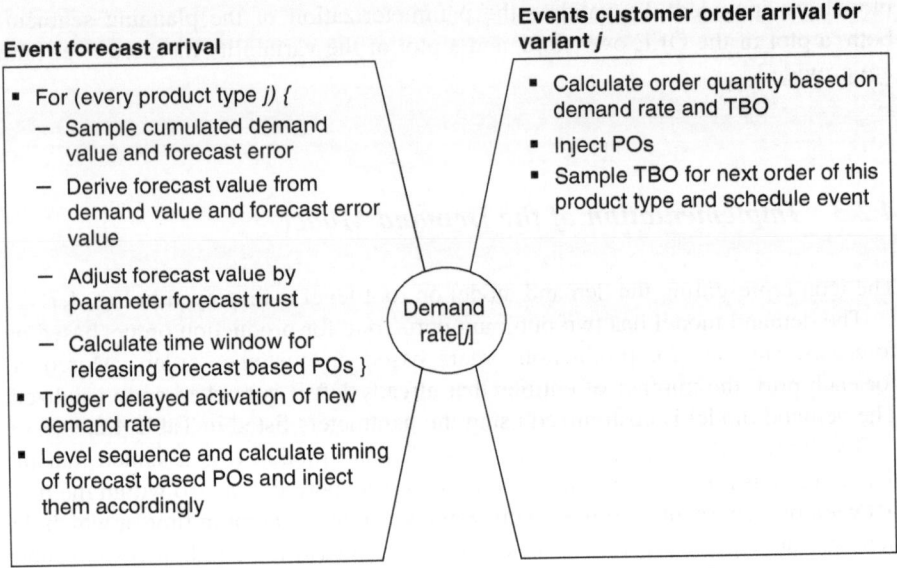

Event forecast arrival

- For (every product type *j*) {
  - Sample cumulated demand value and forecast error
  - Derive forecast value from demand value and forecast error value
  - Adjust forecast value by parameter forecast trust
  - Calculate time window for releasing forecast based POs }
- Trigger delayed activation of new demand rate
- Level sequence and calculate timing of forecast based POs and inject them accordingly

Events customer order arrival for variant *j*

- Calculate order quantity based on demand rate and TBO
- Inject POs
- Sample TBO for next order of this product type and schedule event

Demand rate[*j*]

Fig. 4.8 Pseudo-code for forecast and customer order arrival events

demand rates when calculating the order quantity using the respective equations from the formal demand model Chapter 3.4

The full graphical AnyLogic representation of the demand model can be found in Appendix 8.9.2. The operation of the demand model is mainly driven by two types of events, the forecast arrival (*FCarrival*), and the customer order arrivals for different products (*OrderArrival*). According to the logic defined in Sect. 3.4.1 for the operation of the demand model, the interface between the two events is the current demand rate per product. The logic that is executed if the events occur is described by pseudo-code in Fig. 4.8.

As code example, the Java code for the forecast arrival event (left column in Fig. 4.8) is presented in Appendix 8.9.3.

To enable validation of the parameterization of the demand model, it offers, for each end product, a time plot of the cumulated production orders based on forecasts and based on customer orders. An exemplary time plot (with no forecast error assumed) can be found in Appendix 8.9.2. All demand probability distributions are encoded in a separate function (*Ddist*). Within this function, correlated demands among the products can be realized, also by including third party statistics application programming interfaces (APIs). However, this topic will not be treated further here.

4.2.4   Implementation of the Production System Model

The production system model, which is introduced in the following, utilizes the previously described planning segment and demand model objects in order to map

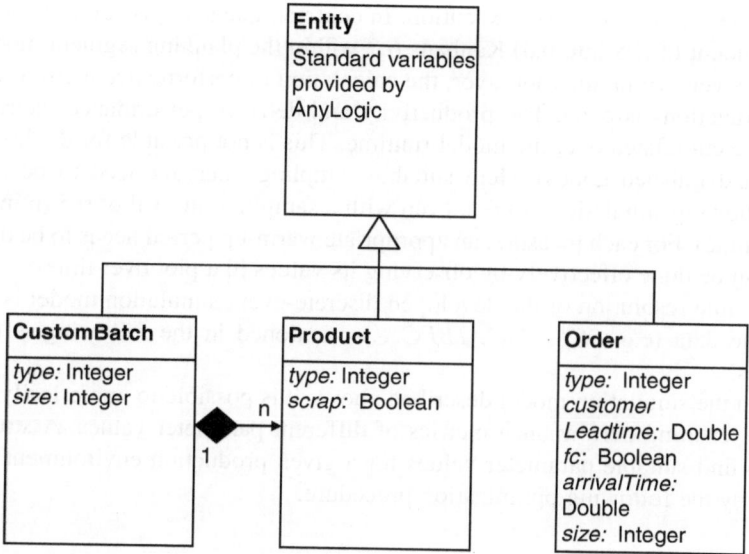

Fig. 4.9 UML diagram of flowing object classes

the full system and its PCS logic. The connections between the objects are implemented according to the queuing network model presented in Chap. 3. In the implementation, also transportation delays between planning segments can be considered. An exemplary a production system model can be found in Appendix 8.11.1.

The objects that flow within the system are instances of different class types. The implemented classes and their relations are summarized in the UML[5] diagram shown in Fig. 4.9.

The custom built flowing objects (*CustomBatch, Product, Order*) are all subclasses of the *Entity*-class provided by AnyLogic. Single products (units) are aggregated into batches represented by the *CustomBatch* class before leaving planning segments. Production orders are instances of class *Order*. A boolean variable (*fc*) indicates whether they are forecast- or customer order-based. All three custom entity classes share the *type* variable, which codifies the product type and is necessary to realize type specific matches as described in Chap. 3 (i.e. *VSSt*). Both, Kanbans between planning segments and the internal Kanbans are not type specific and can therefore be modeled as instances of the *Entity* class.

Whereas the parameter FCT_j has already been integrated in the demand model, the remaining optimization parameters (K_i and S_{ij}) are implemented within the root level by injecting the corresponding number of Kanbans and quantities of basestock

---

[5] Unified modeling language (www.uml.org)

before the start of the model execution. In contrast, queue IC_i is equipped with the right amount of (PS internal) Kanbans (c_i) within the planning segment model.

Moreover, within the root layer, the calculation of performance metrics and the utility functions is done. The productivity and delivery performance metrics can easily be cumulated over the model runtime. This is not possible for the inventory (WIP and finished goods). Here suitable sampling intervals need to be chosen. The following simulations were all run with a sampling interval of 0.5 (minutes in model time). For each measure, an appropriate warm-up period needs to be defined. This can be done effectively by observing its values in a plot over time.

The time resolution of the developed discrete-event simulation model is 1 min. All time data (e.g. SCT, TBO, $TBFC$...) mentioned in the following is thus in minutes.

With the simulation model described above, it is possible to numerically evaluate the resulting performance metrics of different parameter values. A structured way to find suitable parameter values for a given production environment is described by the following optimization procedure.

4.3   A PCS Engineering Process to Optimize Production Control Parameters

4.3.1   Approach and Process Overview

Previous analysis revealed (Sect. 3.2.4) that the large size of the solution space of the proposed queuing network model for PCS engineering prohibits a simple enumeration of all possible solutions and a more sophisticated way to explore it needs to be found. This reflects the complex nature of PCS engineering. In the following, an engineering process that represents an accurate and practically applicable heuristic for defining a suitable PCS is proposed. The approach enables a sequential optimization of certain classes of variables in contrast to optimizing all of them concurrently. This diminishes the combinatorial explosion and enables further ideas to reduce the size of the solution space by delimiting the solution variables and reducing their number.

In order to perform a split optimization of the different classes of optimization variables, potential interactions among them need to be analyzed. Table 4.3 summarizes their interactions.

Table 4.3 Interrelations of optimization variables

| The value of... | ...has an influence on the optimal selection of... | | | 1: true / 0: false |
	K_i	OPP_j	FCT_j / S_{ij}	
K_i		1	1	
OPP_j	0		1	
FCT_j / S_{ij}	0	0		

Like in the EKCS (Dallery and Liberopoulos 2000), and according to the operating curve concept (Nyhuis and Wiendahl 1999), K_i drives the system's capacity and lead-time relevant WIP between the planning segments. Due to this lead time effect, K_i has an influence on the allocation of the OPP and the upstream control approach, no matter if upstream control is organized as MTS, MTF, or hybrid MTF/MTS. The choice of the OPP does not feedback on the choice of K_i though, but on the configuration of the upstream control. Thus, the proposed approach shares the ability of being able to separately optimize K_i and S_{ij} with the EKCS (Dallery and Liberopoulos 2000). Depending on where the OPP is positioned, the upstream control needs to ensure the material supply and hedge against the upstream production system variability. The choice of the upstream control neither has an influence on the optimal choice of K_i, nor on the OPP allocation since the WIP it induces has no influence on system lead time or capacity. From these considerations, it can be concluded that a sequential solution procedure needs to be performed in the logical order (1) K_i, (2) OPP_j, (3) FCT_j and S_{ij}. This stepwise optimization can be interpreted as starting from a world with reduced complexity and only production system variability based drivers, followed by a stepwise integration of demand variability.

The resulting PCS engineering process is depicted in Fig. 4.10.

In step one, an ideal world with no demand uncertainty is assumed. No demand uncertainty is equivalent to full information on demand, which can be interpreted as an infinite customer lead time. Thus, a make-to-order (MTO) approach is indicated. One task in this stage is the definition of structure-driven buffers, which are necessary for instance if equipment is shared for parts that go into the same final product. Moreover, driven by the presence of production system variability and

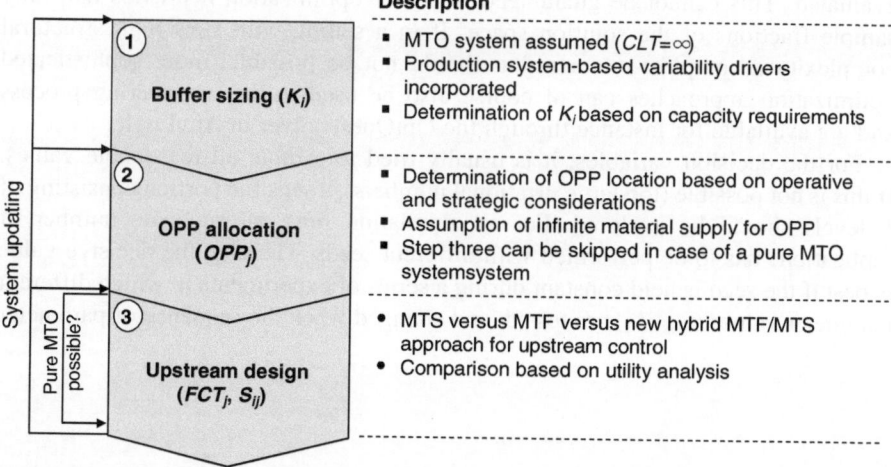

Fig. 4.10 PCS engineering process

capacity requirements, the buffers between consecutive planning segments are sized.

In step two, the assumption of infinite customer lead time is resolved and the OPP needs to be positioned. A major criterion in this step is the comparison of the resulting production lead time from the system designed in step one with the customer lead time. However, also strategic aspects can be taken into account in this step. If for strategic reasons, several OPP options are considered, the following steps can also be continued for each of them and the better option is selected in the end. If the OPP is positioned in front of the first planning segment and the system can be run as pure MTO, the next step can be skipped.

In step three, the approach to deal with demand uncertainty is addressed. It is determined whether a pure MTS or MTF system dominates, or if a hybrid MTF/MTS system as introduced in Chap. 3 is indicated.

For each step, the relevant system parameters that drive the decisions should be determined and on their change, a continuous updating of the configuration be performed. Since the design in each stage builds on results of the previous stages, the whole process needs to be run from the beginning of the stage, in which the parameter that has or will change had an influence for the first time.

The separate optimization as described above reduces the size of the solution space significantly.[6] As we will see in the following, further reductions are possible. Moreover, as we will see in Chap. 5, a relation between FCT_j and S_{ij} exists and their values can be determined analytically, what will further reduce the size of the solution space.

As optimization technique within the three process steps, exhaustive search (Horst and Pardalos 2002) with an adequate number of replications is used. This technique is chosen because it is feasible due to the presented and forthcoming size reductions of the solution space, and because it yields with certainty the best solution from the evaluated options in the search space since all of them are evaluated. This cannot be guaranteed for other optimization heuristics that only sample fractions of the solution space. If in a setting with very high structural complexity a complete enumeration should not be possible, more sophisticated optimization approaches can of course also be used in the engineering process and are available for instance through the OptQuest solver in AnyLogic.

For the decision variables, it is usually tried to sample all reasonable values. If this is not possible (too large, irrational numbers), a sensible portion consisting of L levels should be explored. For all simulation runs, an adequate number of replications has to be performed with different seeds. Thereby, the decisive value is best if the seed is held constant during a series of experiments in which different parameter values (e.g. K_i) are tested, and changed when the sequence of parameter

---

[6] For details see Appendix 8.8

values starts over again (i.e. another replication is started). Moreover, the model runtime needs to be determined appropriately and be cross checked with reality and convergence. A detailed discussion of selecting model runtime, number of replications, and seeds can be found in Law and Kelton (2008).

To ease the following mathematical illustrations of the three optimization steps, a serial material flow is assumed.

4.3.2   Step 1 – K_i Determination

To determine K_i, a saturated version of the queuing network model is examined. Saturated means that an infinite number of orders is induced to be able to evaluate the system's capacity. If the different product types mapped in the model have different standard processing times, the saturated system needs to be loaded with the expected order mix to be as accurate as possible.

In the following, the reaction of the queuing network system to different values of K_i is illustrated with the help of a simple three-stage serial production system (see Appendix 8.10.1 for its parameterization). The left graph in Fig. 4.11 shows how the productivity of the whole production changes if K_1 is varied and K_2 is held constant at a high level. The second graph shows the respective variation of K_2.

Productivity (PROD) is calculated as percentage of the maximum achievable throughput in a perfectly line balanced system with no disturbances. Therefore, the produced quantities (end products) of each variant are weighted with the sum of their standard cycle times SCT and summed up. This sum is then divided by the considered time period times the number of planning segments (4.5). The quotient equals 1 if all planning segments were able to produce without disturbance during

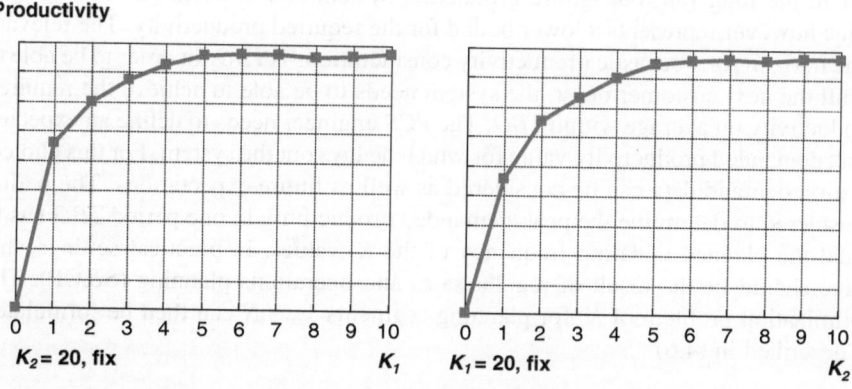

Fig. 4.11 Examples of numerically determined operating curves

the whole time period (what implies that the system is already filled at the start of
the considered time period).

$$PROD = \frac{\sum_{j=1}^{M} \left(\sum_{i=1}^{N} SCT_{i,j} \right) \cdot X_j}{timePeriod \cdot N} \in [0; 1] \qquad (4.5)$$

timePeriod: Time period for which the productivity is calculated
       N: Number of planning segments
    SCT_{ij}: Standard cycle time, i.e. time to produce one unit of product j in
            planning segment i if no disturbance occurs (equals processing time
            divided by the number of units processed simultaneously)
      X_j: Cumulated amount of finished goods j produced during *timePeriod*

The numerically determined curves exhibit the typical shape of operating curves
as introduced by Nyhuis and Wiendahl (1999). The increase in productivity
decreases with higher K_i and converges against a maximum. These variations of
one parameter should be used to screen for reasonable upper and lower bounds
(k_i^{lower}, k_i^{upper}) within which the parameters K_i will be ranged during their simulta-
neous variation. Establishing these boundaries again can reduce the solution space.

The optimization problem is to determine K_i such that the resulting system
capacity is just sufficient to satisfy demand with lowest possible WIP. Thus, it is
assumed that it is always desirable to produce an additional unit to satisfy demand.
Given that the production system is dimensioned correctly, this assumption should
be valid in practical applications. The driving factors in this optimization step are
obviously the demand, the standard processing times, and all factors that lead to
an increased actual cycle time (i.e. OEE losses) as described in Sect. 3.3.

A simple but misleading approach would be to set K_i such that the expected
value of the production system's productivity equals the expected value of the
productivity as required to fulfill demand. This would ensure that demand can be
met in the long run, but ignore production system and demand variability. This
value however represents a lower bound for the required productivity. The relevant
time horizon for stochastic productivity considerations is *TBO*. In order to be able to
fulfill the next customer order, the system needs to be able to achieve the required
productivity on average within *TBO*. The PCS engineer needs to define an expected
peak demanded productivity value for which he lays out the system. For this choice,
historic demand data can be considered as well as future expectations. The period
considered to determine the peak demanded productivity in one period *TBO* has to
equal the planned updating frequency of the K_i values. In practical settings, this
value should be the result of regular sales and operations planning (S&OP). The
optimization problem of K_i for planning segments $1, \ldots, N$ can then be formulated
as described in (4.6)

$$\min \ WIP(K_1, ..., K_N)$$

$$P(PROD_{TBO}(K_1, ..., K_N) \geq prod_{TBO}^{DemandedPeak}) \geq p$$

$$s.t. \quad E(PROD_{TBO}(K_1, ..., K_N)) \geq prod^{DemandedAvg} \tag{4.6}$$

$$k_i^{lower} \leq K_i \leq k_i^{upper} \quad \forall i = 1, ..., N$$

$$K_i \in \mathbb{N}$$

$WIP(K_1, ..., K_N)$:   Numerically determined function representing the average WIP in simulation runs resulting from values K_i (deterministic value)

$PROD_{TBO}(K_1, ..., K_N)$:   Numerically determined productivity levels the saturated production system achieves with values K_i in periods TBO (set of achieved productivity levels in simulation interpreted as random variable)

$E(.)$:   Expected value (approximated by sample mean)

$P(X \geq x)$:   Probability that random variable X reaches a value larger than or equal to x (approximated over sample)

$prod_{TBO}^{DemandedPeak}$:   Required productivity level to satisfy peak demand in relevant time period TBO

$prod^{DemandedAvg}$:   Required productivity level to satisfy demand on average

k_i^{lower}, k_i^{upper}:   Upper and lower bounds for K_i as determined in single factor variation screening

p: p-value $\in [0,1]$ for the stochastic optimization; probability with which the peak volume is achieved in TBO

Thereby, the values for K_i will be tested on a reasonable number of L levels, which are uniformly distributed, ideally with the minimal step size 1. By using the expected average demand, the second constraint forms a lower bound for the required productivity. If the achieved average productivity is smaller than the demanded average productivity, the system would accumulate backlog. However, matching the averages over time is not sufficient since the system needs to be able to deliver on time. Therefore, the potentially stricter first constraint is introduced which requires a certain probability (parameter p) to achieve peak productivity between two consecutive orders. If the p-value is set to zero ($p = 0$), the optimization yields the lowest reasonable bound for K_i. Setting $p = 1$ leads to the upper reasonable bound. Further increasing K_i only leads to more WIP but no productivity improvement noticeable within the relevant time horizon TBO. The remaining question is now finding a reasonable value for p. If $p < 1$, the probability to not achieve the required capacity in one order period is $1-p$. Depending on whether the planning segments are located upstream or downstream the OPP, this effect could be compensated by the choices in steps two and three. Thus, there is a trade-off that needs to be considered further.

In the upstream part, the potentially delayed arrival of material could be compensated by increasing S_{ij}, (or, in the MTF case, $FCadvance$). For the EKCS, which is connatural in this aspect, Liberopoulos and Koukoumialos (2005) performed an analysis on the trade-offs between K_i and S_{ij}. Their results suggest that it is always more cost efficient to cover production system variability with K_i instead of with more basestock since basestock needs to be held for each variant. Moreover, they show that increasing K_i even more than suggested by $p = 1$ does not change the proposed value of S_{ij}. In our case, their second statement is true for the value of a pure MTS system, however, it is found that the over capacity influences under certain assumptions the later considered decision among MTS and MTF (see analysis in Appendix 8.10.2).

For the downstream part, the risk of not achieving the needed capacity can be hedged against by allocating the OPP further downstream than it would be optimal with near certain capacity achievement. However, this has the drawback that the variability the upstream control needs to cover increases (more planning segments, longer lead time) and thus, WIP increases here again for each variant. In addition, the effectiveness of this hedge depends on the existence and location of a bottleneck in the downstream part.

We therefore propose to run the optimization with a high p-value (0.99 or the highest achievable value in case of an over-loaded system). In most cases, this will already lead to the optimal value. To back it up against the mentioned trade-off downstream the OPP, a sensitivity analysis with the results of lower p-values can be performed in engineering process step two. If a lower p-value should lead to lower K_i's that then lead to another feasible solution for the OPP further upstream, for this option, optimization step three can also be performed and the better solution be selected in the end by comparing utility values. However, in the experiments conducted with the PCS engineering framework so far, this case has not yet been observed.

The described proceeding is in line with the thinking propagated in Lean Manufacturing. There, limiting and minimizing WIP between process steps in order to ensure a short production lead time is key since it leads to many follow-on benefits like shorter quality feedback loops. In addition, system improvements are stimulated with this approach since no unnecessary capacity contingencies are allowed that would cover-up disturbances. To make inefficiencies even more visible, K_i can be reduced slightly below the optimum so that disturbances cause immediate 'pain' in the organization and hence increase the improvement pressure. This is one of the core principles in Lean Manufacturing. Thus, by varying p, production management can control the operational risk on the one hand, and improvement speed of the plant on the other hand.

4.3.3   Step 2 – OPP_j Determination

A survey of different strategic and operative considerations on positioning the OPP can be found in Sect. 2.1.3 and Olhager (2003). In the following, only the quantitative considerations of OPP positioning are presented. From the viewpoint of our queuing network model, it is optimal to allocate the OPP as far upstream as permitted by the customer lead time. The reason behind this is that the more upstream the OPP is, the less lead time and production system variability needs to be covered by the stocks in front of the OPP since it is covered, free of cost, by the available customer lead time. Thus, the whole system is able to operate with minimum WIP.

For the following optimization, it is assumed that in front of the OPP an unlimited material supply is available. The system is being engineered to meet customer expectations, thus the targeted delivery performance from the OPP downstream is $delPerf_{upperBound}$ as introduced in Sect. 4.1.2. The trade-off among delivery performance and the WIP level driven by upstream control will then be made in optimization step three. The optimization can be done separately for each product j. The optimization problem is given by (4.7)

$$
\begin{aligned}
\min \quad & OPP_j \\
s.t. \quad & DelPerf(OPP_j) \geq delPerf_{upperbound} \\
& OPP_j \in \{1, ..., N+1\}
\end{aligned}
\tag{4.7}
$$

$DelPerf(OPP_j)$:   Numerically determined delivery performance in simulation runs with OPP located in front of planning segment OPP_j

$delPerf_{upperBound}$:   Upper bound delivery performance as expected from the customers/market (see value function Sect. 4.1.2)

How far the OPP can be positioned upstream depends also on the chosen values of K_i. As mentioned before, the OPP allocation optimization can be done for several p-values in the K_i optimization in order to check the sensitivity. If the unlikely case occurs that for a p-value lower than 0.99 the OPP can be moved further upstream and still satisfies the delivery performance condition, this scenario should be treated as an option and step three be as well performed for it. Also if for strategic reasons, an optional position of the OPP further downstream than recommended by the optimization is considered, this position can be carried on into optimization step three. The best option is then selected in the end by comparing utility values.

We mentioned that the variants can be optimized one by one in this optimization step what further reduces the size of the solution space.[7]

4.3.4   Step 3 – FCT_j and S_{ij} Determination

For the planning segments upstream the OPP, the approach to deal with demand variability needs to be defined in order to be able to timely provide the planning

---

[7] For details see Appendix 8.8

Table 4.4 Parameterization of the value function for the basestock allocation experiment

Dimension	Weight	Parameter	Value
WIP	0.30	Minimum WIP	50
		Maximum WIP	350
Delivery performance	0.70	Lower bound	0.70
		Upper bound	0.95

segments after the OPP with the required material. The presented PCS engineering framework offers the options MTF, MTS, or hybrid MTF/MTS, depending on the choice of parameters FCT_j and S_{ij}.

At this point, the trade-off between delivery performance and WIP is made and a utility function according to Sect. 4.1.2 needs to be defined. Appendix 8.10.3 shows how utility improvements, which play an important role in the following, can be converted into potential WIP reductions to make them more tangible.

In the proposed queuing network model, upstream the OPP, in front of each planning segment, basestock S_{ij} can be positioned. However, to reduce the size of the solution space further, and due to the challenges implementing this spread basestock in practice would bear, a general analysis on the optimal base stock location for a product j has been performed. The hypothesis for the following analysis is that under certain conditions, it is optimal to allocate all basestock right in front of the OPP. The applied utility function for the experiment is summarized in Table 4.4. The full experimental setup can be found in Appendix 8.10.1.

Sixteen basestock levels ([0,...,15]) were tested at two possible basestock locations in a three-stage serial production system; one directly in front of the OPP which is located at the end of the line (S_{4j}) and one in front of the second planning segment (S_{2j}). The results of the complete enumeration are displayed in Fig. 4.12. The numerically evaluated configurations can be read from the graph in the lower right section of Fig. 4.12. The resulting value of each configuration is displayed in the graph in the upper left corner. The underlying values of achieved delivery performance and WIP are presented in the graph in the upper right corner.

The numerically determined optimum is $S_{2j} = 0$ and $S_{4j} = 12$. This result is plausible due to the following logic. The hedging effect against upstream production system variability of one piece of WIP is largest right in front of the OPP since then, the number of covered planning segments by one piece of basestock is maximized. However, this is only applicable if the model assumes that the WIP has equal value from a cash-flow point of view, no matter where it is positioned. Whenever this assumption can be made, it is valid to consider positioning of basestock right in front of the OPP only. If not, a WIP evaluation function needs to be defined and basestock be optimized in all locations $i \leq OPP_j$. The WIP valuation function for each stage can be reflected in the value function used for WIP according to Sect. 4.1.2. Besides different values of the WIP, another scenario would justify a deviation from the rule proposed above. If two or more products share parts and the demands of these products are correlated, a risk-pooling effect can occur and enable a further WIP reduction. We will address this scenario in more detail in Sect. 5.2.3

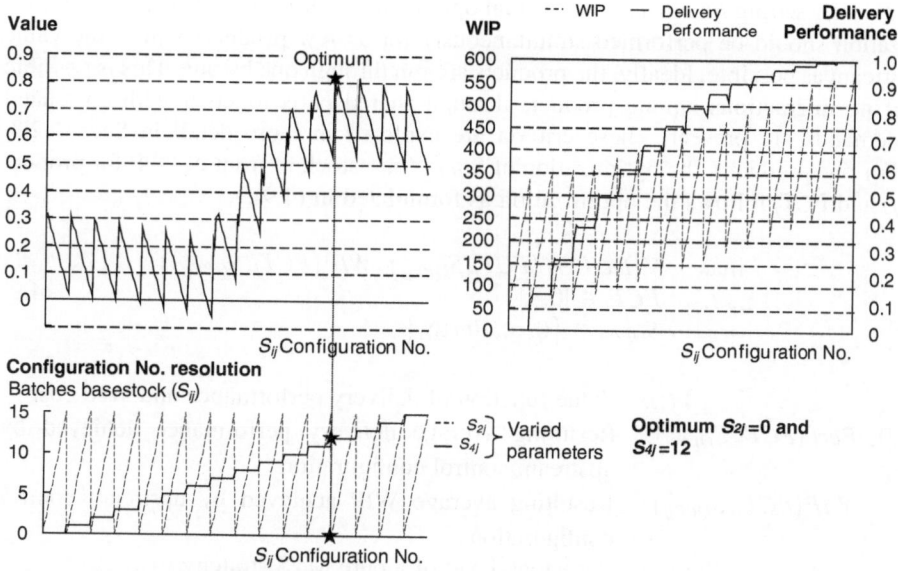

Fig. 4.12 Basestock allocation experiment

Considering basestock only in front of the OPP would again reduce the solution space.[8]

Before running the main optimization, the MRP parameter *FCadvance* needs to be set according to the following optimization (4.8).

$$\max \quad V[DelPerf(FCadvance), WIP(FCadvance)]$$

$$FCerr_j = 0 \quad \forall j = 1, ..., M$$

$$s.t. \quad FCT_j = 1 \quad \forall j = 1, ..., M \tag{4.8}$$

$$FCadvance \in [0, FCarrivalAdv]$$

$V(.)$:   Value function of delivery performance and WIP cost
$DelPerf(FCadvance)$:   Resulting average delivery performance achieved by *FCadvance*
$WIP(FCadvance)$:   Resulting average WIP achieved by *FCadvance*
$FCarrivalAdv$:   Maximum possible forecast advance, depending on how much in advance forecast is received

The optimization of *FCadvance* is performed under the assumption of no forecast error and full forecast trust. Again, a reasonable amount of levels to be tested should be selected since not all values in the interval [0,*FCarrivalAdv*] can be probed.

---

[8] For details see Appendix 8.8

After setting *FCadvance*, the actual optimization can be performed. The optimization should be performed simultaneously for as few products within the value stream as possible. Ideally, the products are run through one by one. This is possible if no interactions among products through shared parts together with correlated demands are present. These criteria are explored in more detail in Sect. 5.2.3. For one product j, the optimization of $S_{OPP,j}$ (basestock in front of OPP for product j) and FCT_j follows the optimization as formulated in (4.9).

$$
\begin{aligned}
\max \quad & V\left[DelPerf\left(FCT_j, S_{OPPjj}\right), WIP\left(FCT_j, S_{OPPjj}\right)\right] \\
s.t. \quad & FCT_j \in [0, 1] \\
& S_{OPPjj} \in \{0, ..., PureS_j\}
\end{aligned}
\tag{4.9}
$$

$V(.)$:   Value function of delivery performance and WIP cost

$DelPerf\left(FCT_j, S_{OPPjj}\right)$:   Resulting average delivery performance achieved by upstream control configuration

$WIP\left(FCT_j, S_{OPPjj}\right)$:   Resulting average WIP achieved by upstream control configuration

$PureS_j$:   Basestock level of a pure MTS strategy

It is sensible to upfront scale the search space for S_{OPPjj} by identifying the basestock level $PureS_j$ of a pure MTS strategy ($FCT_j = 0$) through simulation. $PureS_j$ is an upper bound for S_{ij}. If the search space is to large, a reasonable value for the number of tested levels L and an according step size have to be chosen (for example, use step size 2). The concurrent optimization of several products is, where necessary, performed in the same manner by maximizing the sum of values over the products that are considered simultaneously. Figure 4.13 shows an exemplary optimization plot of the example described in Appendix 8.10.1. Again, a complete enumeration of all configurations of FCT_j and S_{ij} is performed. Like in the basestock allocation experiment, the lower right graph nominates the tested configurations. Above, graphs show the resulting value and the underlying realizations of delivery performance and WIP of each configuration. For this illustration, the same utility function as in the basestock allocation experiment (Table 4.4) is used.

The optimization run yields an optimum at $FCT_j = 0.7$ and $S_{OPPj \; j} = 7$. This means that in this setup, a hybrid MTF/MTS strategy dominates and outperforms the pure MTS ($FCT_j = 0$ and $S_{OPPjj} = 15$) and pure MTF ($FCT_j = 1$ and $S_{OPPjj} = 0$) strategies.

If preferably all products can be optimized separately according to the conditions mentioned above, the maximum reduction of the solution space increases further.[9]

---

[9] For details see Appendix 8.8

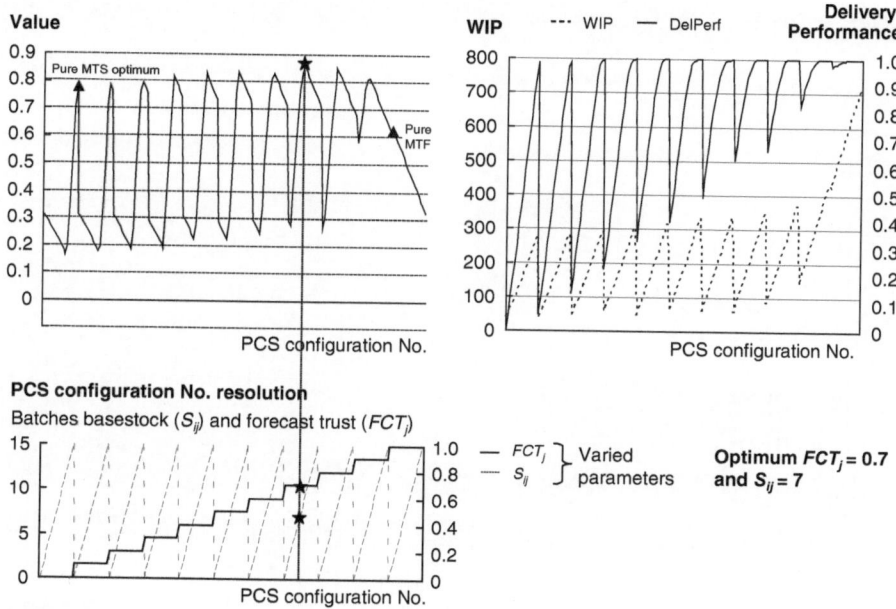

Fig. 4.13 Example for FCT_j and S_{ij} optimization run

To sum it up, a PCS engineering framework for the three key questions in PCS engineering in complex discrete manufacturing is provided. The physics behind the ideas of step one and step two of the proposed optimization process have already been addressed in detail by Nyhuis and Wiendahl (1999) and Olhager (2003) and are obvious from the discussions in this chapter. For step one, they are based on comparing the required production capacity with the available capacity that is subject by the production system variability. In step two, the major driver is the comparison of the customer lead time with the production system lead time, which is driven by the processing times, the production system variability, and the WIP levels based on the K_i identified in step one. However, the physics behind step three and the enabled hybrid strategies are not yet clear and bear an unexploited improvement potential. Therefore, the following exploration of the queuing network model based PCS engineering framework will have its focus on the upstream control strategies (step three) and the hybrid MTF/MTS (Push/Pull) approach.

Chapter 5
Investigation of the Push/Pull Integration

5.1 Influencing Factors and Rules for the Decision among Push and Pull

5.1.1 Hypothesis and Experiment Design

In the following, relevant drivers for the decision among push (referred to as pure MTF) and pull (referred to as pure MTS) will be identified by structured experiments, in which different operating conditions are varied. Based on this, decision rules for the application of pure MTS, pure MTS and the hybrid MTF/MTS will be derived.

For the choices of parameters S_{ij} and FCT_j, which determine the upstream control mode, the hypothesis in the following is that they are, just like parameters K_i and OPP_j, driven by a definable group of design drivers and that a basic decision rule to chose between pure MTS, pure MTF, and hybrid MTF/MTS can be found.

To investigate the driving forces, a structured series of experiments is employed in which the optimal solution is determined using the PCS engineering process for different environmental settings. A serial three stage production system with one product is used as basic experimental setup. The AnyLogic production system model representing it, which will also be used for all other experiments in Chap. 5, can be found in Appendix 8.11.1. The cash-flow value of WIP can be assumed to be constant over all planning segments. The experiment utilizes a fractional facto-rial experimental design[1] with five factors varied over two levels. Varying factors over two levels reveals linear relations between the factors and the response variable, but would not discover non-linearities. This is sufficient here since the main objective is sorting out relevant factors for which more detailed analyses follow. A fractional-factorial design is chosen to reduce the necessary number of

---

[1] For an introduction and the theoretical background of experiment design and analysis, it is referred to Montgomery (2009)

C. Karrer, *Engineering Production Control Strategies*, Management for Professionals,
DOI 10.1007/978-3-642-24142-0_5, © Springer-Verlag Berlin Heidelberg 2012

simulation runs. The chosen design provides a resolution level V. This means that "no main effect or two-factor interaction is aliased with any other main effect or two-factor interaction, but two-factor interactions are aliased with three factor interactions" (Montgomery 2009). This is sufficient for the purpose of screening factors. Together with three replications for each factor combination, the resolution V design requires 48 simulation runs in total, compared to 96 runs,[2] which a full factorial design would require. The detailed system parameterization and the high and low values for the five varied factors are care documented in Appendix 8.11.2. The choice of factors and levels will be explained in the following.

From the available PCS design drivers, five factors that might have an impact on the definition of an upstream control strategy were chosen to be part of the fractional factorial experimental design. In the planning segment section, the breakdown time to repair (TTR) and the batch size are varied. The remaining variability drivers would as well affect the process variability measure ($VMPS$), however, the targeted variability change can also be induced by varying only one factor. Also the line length is not changed since the PCS relevant effect of a longer line would be the same as increasing the process variability directly in the planning segments. In the demand model, the customer lead time can be ignored since it has no effect on the choice of control upstream the OPP, which is the focus of this experiment. Moreover, $TBFC$ is held constant. Here, the deciding point is its relation to the TBO, which is varied. The number of products is constant (one). The extension to the multi-product case will be discussed later in Sect. 5.2.3.

For each chosen factor, a high and a low value is assigned. Thereby, the values are chosen from the extreme ends of their ranges in order to show a potential effect as clear as possible. However, their values are set such that they are still reasonable for complex discrete manufacturing systems and physically sensible. The low value for forecast error is represented by no forecast error (0) and a resulting coefficient of variation (CV) of the forecast error distribution of 0. The high value is modeled by a uniformly distributed forecast error that can reach values up to 0.8 and yields a CV of 0.46. Low demand variability is represented by a constant demand, high demand variability by a demand pattern with a low base load and occasional high peaks. It is represented using a Bernoulli distribution, leading to a CV of 1.53. The high value of TBO is set to $\frac{TBFC}{2} = 5000$ and the low value to 1,000, which means an order every 16.7 h.[3] The order and transport batch sizes are assumed to be synchronized. The maximum batch size of 40 is chosen such that at least two batches are produced during a forecasting period. The minimum batch size is set to 10, which corresponds to a low average standard processing time of a batch of 18.3 min.[4] For the process variability, either a constant time to repair for breakdowns is

---

[2] A full factorial design for 5 factors varied on 2 levels with 3 replications would need $2^5 \cdot 3 = 96$

[3] $\frac{1000 \, min}{60 \frac{min}{h}} = 16.7h$

[4] $\frac{1}{N} \sum_{i=1}^{N} SCT_i \cdot BatchSize$

assumed, or a uniformly distributed time that can range between 0 and 300. This leads to respective *VMPS* measures of 16,700 and 22,700.

Each simulation run lasts for 600,000 min (roughly 1 year in the assumed shift model).

For the following experiments and comparisons, a utility function to evaluate delivery performance and WIP resulting from the chosen upstream control approach needs to be assumed. For the choice of the weights and the bounds of the delivery performance, extreme values were avoided to get to a function that approximately represents, based on the experience of the author, the valuations of an 'average' company. The WIP limits are based on the observations made in the different simulation runs. Thus, it is implicitly assumed that there are no physical limits that constrain WIP within the observed range. It is further assumed that the conditions for the application of an additive utility function as presented in Sect. 4.1.2 are met. The parameterization of the function is summarized in Table 5.1. The same function is applied for the remainder of this chapter.

For each simulation run, result measures will be computed for later analysis. Table 5.2 summarizes them. They are not only used for the following experiment, but for all experiments that will follow within this chapter.

Next, the analysis of the results from executing the described experiments will be presented.

Table 5.1 Parameterization of the value function for the following experiments

Dimension	Weight	Parameter	Value
WIP	0.30	Minimum WIP	50
		Maximum WIP	350
Delivery Performance	0.70	Lower bound	0.70
		Upper bound	0.95

Table 5.2 Summary of result measures used in experiments

Result measure	Unit	Formula	Description
FCT*	Percent	N/A	Optimal forecast trust value (here FCT_1)
S*	Batches	N/A	Optimal basestock level (here for S_{41})
Sreach*	Time units	$\frac{S^* \cdot Batchsize}{PeakDemand} \cdot TBFC$	Time supply of optimal basestock level under peak demand
PureS	Batches	N/A	Optimal basestock level in a pure MTS system ($FCT_j = 0$)
S%*	Percent	$\frac{S^*}{PureS}$	Optimal basestock level expressed as percentage of the basestock level of pure MTS
Utility*	$\in [0, 1]$	N/A	Utility achieved with FCT* and S*
UtilityMTF/ UtilityMTS	$\in [0, 1]$	N/A	Utility of pure MTF or pure MTS systems
UtilityImprMTF	Percent	$\frac{Utility^* - UtilityMTF}{UtilityMTF}$	Percentage utility increase achieved by optimal solution compared to pure MTF
UtilityImprMTS	Percent	$\frac{Utility^* - UtilityMTS}{UtilityMTS}$	Percentage utility increase achieved by optimal solution compared to pure MTS

5.1.2 Determination of Relevant Factors and Derivation of Decision Rules

The full experimental configurations as described in the previous section and the corresponding simulation results on which the following analysis builds can be found in Appendix 8.11.2. To analyze the results, a general linear model is built as described in the appendix. For this and all following statistical analysis, the software MiniTab™ (Appendix 8.13) is used.

The effects of the varied parameters are displayed as "normal probability plot of the effects" (Montgomery 2009). In this graphical representation, negligible effects are normally distributed with mean zero and form approximately a straight line. Relevant factors fall outside the line (Montgomery 2009).

The experiment clearly shows that the driving factors for the choice of FCT^* are the demand variability (*Demand_var*), the forecast error (*FC_err*), and their interaction effect. The other factors are irrelevant for the choice of FCT^* (see Fig. 5.1).

Process variability (*Process_var*) has no influence on the choice of FCT^* since in a system with sufficient capacity (which is the case in this experiment), it is entirely covered by the choice of K_i. In a system that operates at its capacity limit, process variability can have an impact on the optimal MTF/MTS induced stock level, which then also starts hedging for production system variability, however, less efficient (see argumentation in Sect. 4.3.2). Nevertheless, also the choice between MTS and MTF is not affected. The time between order (*TBO*) and the batch size (*Batchsize*) have, according to the analysis above, no influence on the choice between MTS and MTF. However, looking at the result data (in the appendix) more closely, it suggests that *TBO* and the batch size might have an

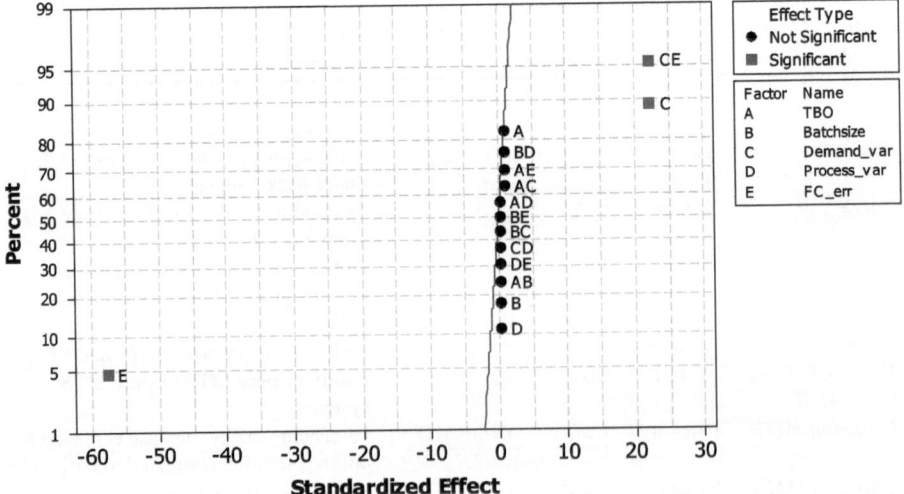

Fig. 5.1 Normal probability plot of standardized effects for FCT^* (Alpha = 0.05)

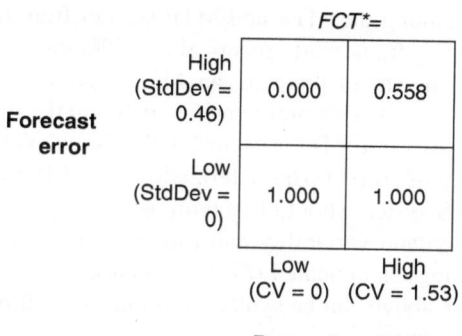

Fig. 5.2 Cube plot (data means) for *FCT** (StdDev: Standard deviation, CV: Coefficient of variation)

impact on the magnitude in which the choice of *FCT* influences system performance. This will be examined later in Sect. 5.3.

To better understand the effect of demand variability and forecast error on the choice of *FCT*, a 'cube plot' is constructed. The cube plot shows for all combinations of the influencing factors the resulting average value of *FCT** (Fig. 5.2).

In the lower right corner, demand variability is present but no forecast error. Here, a pure MTF system (*FCT** = 1) is recommended. Under the assumption of a correctly parameterized MRP system, this result is coherent since whereas a pure MTS system would always have sufficient stock in order to satisfy peak demand, an MTF system only generates the right amount of stock (no forecast error assumption) before an actual demand event.

In the upper left corner, no demand variability is present but forecast error. Here, a pure MTS system (*FCT** = 0) is recommended since the just mentioned advantage of MTF vanishes with a constant, non-variable demand and at the same time, MTF has to deal with problems of delivery performance or excess stocks that are caused by forecast errors. This scenario will be important within following discussions, however, it is of theoretical nature since one could argue that in practice, the constant demand rate could just be used as forecast value. This leads to the lower left corner, a situation with neither demand variability nor forecast error. Here, a pure MTF system is recommended (*FCT** = 1). However, this recommendation is only based on a very small average utility improvement of 0.03 (average taken from result data in Appendix 8.11.2) compared to a pure MTS system.

In contrast, the recommendation of MTF in the lower right corner is based on an average utility increase of 0.17 (average taken from result data in Appendix 8.11.2) compared to MTS. For the case with no demand variability and no forecast error, this suggests rather indifference among MTS and MTF, which is underpinned by the fact that for the experiment above, K_i were chosen such that the system has a slight overcapacity. Together with the considerations that were already presented in Appendix 8.10.2, the slight preference towards the MTF system can easily be explained.

Being indifferent among the MTS and MTF system from the standpoint of the present PCS engineering framework, practical considerations should be pulled up. It is commonly known that in practice, an MTS system is inherently easier to control than an MTF system. This preference is rooted in the focus of MTS systems on tangible WIP in contrast to MTF systems, which focus on intangible lead times and bear the challenge of correct parameterization. Thus, if no other reasons speak against it, a pure MTS system should be favored.

Finally, if both, demand variability and forecast error are present (upper right corner), a hybrid strategy is indicated ($FCT^* = 0.58$).

The considerations above can be synthesized into the following three rules:

Rule 1: If there is no demand variability, a pure MTS strategy is indicated.
Rule 2: If there is demand variability but no forecast error, a pure MTF strategy is indicated.
Rule 3: If demand variability and forecast error are present, a hybrid MTF/MTS strategy is indicated.

FCT^* can be interpreted as optimal MTF application percentage. Equivalently, $S\%^*$ can be viewed as optimal MTS application percentage. Therefore, the analyses above were also performed on $S\%^*$, which led to an equivalent result (not shown). Inspired by this result, a regression analysis has been performed on $S\%^*$ and FCT^* (see Appendix 8.11.2), which clearly indicates ($R\text{-}Sq = 94.6\%$, regression p-value $= 0$) that a relation among the two variables exists. This finding will be used as a hypothesis in the next chapter.

5.2 Closed-Form Determination of the Push/Pull Integration Parameters FCT^* and $S\%^*$

5.2.1 Hypothesis and Experiment Design

The objective of the following experiments is to understand the choice of FCT^* and $S\%^*$ (and together with *PureS*, S^* respectively) in more detail. The hypothesis is that a closed-from expression can be found for setting the two parameters and, moreover that a relation between them can be formulated. Having a closed-form expression for FCT^* and $S\%^*$ would significantly ease the application of the hybrid MTF/MTS control approach in practice since the minimal size of the solution space that needs to be screened by simulation is reduced further.[5]

The analytical determination of FCT^* and $S\%^*$ would reduce the size of the solution space by $L \cdot M$ since it would not be necessary to numerically check for M products L levels of FCT^* and L levels of basestock. However, still the basestock

---

[5] For details see Appendix 8.8

level of the pure MTS strategy (*PureS*) needs to be known to compute S^*. If in practice, a pure MTS system with trusted *PureS* values is already in place, even this step could be removed and simulation completely omitted to determine upstream control.

In the following experimental setup, it is assumed that a hybrid MTF/MTS strategy (also referred to as hybrid strategy) is indicated, which means that both, forecast error and demand variability are present. The other factors for which the previous experiment showed that they do not influence the choice of FCT^* and $S\%^*$ are held constant. Based on the results of the first experiment, the values of the constant factors were fixed at either their high or low value such that the utility improvement effect of the hybrid strategy is as large as possible. The details on the magnitude of improvements achievable with the hybrid strategy will be discussed later in Sect. 5.3. For the forecast error, a center point is introduced to be able to detect non-linear relations. Moreover, a hypothetical distribution based on the Bernoulli distribution has been constructed for the forecast error to simplify the discovery of possible quantitative relations involving it. The suggested forecast error distribution takes with equal probability (0.5) either a high value (*a*), or the same value with negative sign as low value ($-a$)

$$- a + bernoulli(0.5) \cdot 2a \tag{5.1}$$

$a \in [0, 1]$ Factor determining the magnitude of the forecast error

This forecast error distribution is unbiased $E(FCerr) = 0$ as demanded by (28). Moreover, its median is also zero $q_{0.5}(FCerr) = 0$, a property which we will discover to be helpful later. To *a* will be referred to as 'expected deviation' in the following.

For the demand, two distributions with increasing coefficients of variation (*CV*) are chosen. One based on uniform distribution, and one constructed based on the exponential distribution to depict a demand pattern with a low base volume and occasional peaks.

All other parameters remain fixed at the high or low value from the previous experiment. The experimental setup is summarized in Appendix 8.11.3.

Due to the limited number of varied factors, a full factorial design can be chosen. This results, for one factor varied over three levels (forecast error), one factor varied over two levels (demand variability), and with three replications per configuration, in 18 simulation runs.[6] The runtime of each replication run is increased to 1,200,000 (min) since a higher accuracy is needed in order to discover analytical relationships compared to the objectives of a factor screening experiment.

---

[6] Number of simulation runs calculated by $2 \cdot 3 \cdot 3 = 18$

For each configuration, the values of K_i are chosen such that the needed capacity is matched, unlike in the previous experiment, where K_i were chosen with slight overcapacity to ease their optimization.

The same result variables as in previous experiment are tracked (see Table 5.2).

5.2.2 Derivation of a Closed-Form Parameterization for Hybrid Control

The result table of the full factorial experiment described above can be found in Appendix 8.11.3.

The experiment has been performed under conditions in which a hybrid strategy is indicated. Based on the results from the experiment in Sect. 5.1, this means that demand variability and forecast error have to be present. We now examine whether the optimal values of FCT^*, $S\%^*$, and S^* are also driven by both factors, or only by one. The hypothesis for this analysis would be that FCT^* and $S\%^*$ are only driven by the forecast error, and demand variability just needs to be present, even though its characteristic does not influence the choice of optimal values. Therefore, Table 5.3 summarizes the p-values for the relevance of each factor and their interaction effect on the choice of several response variables. The details of the statistical analysis, including the residual plots (which show no abnormalities), can be found in Appendix 8.11.3. For the interpretation of the p-values in the table, we use (as for the rest of this chapter) the regular significance level 0.05.

First, the analysis clearly confirms the hypothesis (with significance level 0.05) that the choice of FCT^* and $S\%^*$ is only driven by the forecast error. In contrast to this, it is worth noting that the absolute value of the optimal basestock level, S^*, depends on both, forecast error and demand variability. This can be explained since in the hybrid case, the absolute value S^* depends on the one hand, like the stock level in a pure MTS system, on the demand variability. On the other hand, in the hybrid case, where we found a correlation between the values of FCT^* and $S\%^*$, it can be concluded that S^* is somehow driven by FCT^*, which again depends on the forecast error.

Using the observations above and the decision rules derived in the previous Sect. 5.1.2, a hypothesis for the calculation of FCT^* can be developed. Due to the way the probability distribution for the forecast error is defined in this experiment (5.1), for each product, we can sharply separate the demand into a certain part (with

Table 5.3 Drivers for the optimal value of FCT*, S%*, and S*

Factor	p-values		
	FCT^*	$S\%^*$	S^*
Demand variability	0.29	0.10	0.00
Forecast error	0.00	0.00	0.00
Demand variability*Forecast error (interaction effect)	0.70	0.63	0.17

zero forecast error) and a completely uncertain part, driven by parameter a in (5.1). Using the decision rules from the previous section, we would use a pure MTF for the certain part since there is no forecast error, but demand variability. For the uncertain part, we would use an MTS approach as no forecast data is available for it. Since the forecast error is measured with the actual demand as basis, the certain part of the demand can be separated by multiplying the forecast value with the factor $\frac{1}{1+a}$.

Taking a more generic and probabilistic view, the following generalization to arbitrary forecast error distributions can be made. To realize this idea with as easy as possible mathematical computations in the following and later in practical application, the absolute value function is applied and the earlier introduced assumption $q_{0.5}(FCerr) = 0$ is leveraged. For the forecast error distribution defined earlier for this experiment, (5.2) holds.

$$f(x) = -a + bernoulli(0.5) \cdot 2a \Rightarrow f(x) \in \{-a, a\} \Rightarrow E(|f(x)|) = a \qquad (5.2)$$

Based on this, the following hypothesis for deriving FCT^* can be proposed

$$FCT^* = \frac{1}{1 + E(|FCerr|)} \qquad (5.3)$$

Following the same logic, two hypothesis for determining S^* can be set up. For the uncertain portion of the demand, a pure MTS approach is used. The necessary basestock quantity is the product of the uncertain demand portion and the basestock level of the pure MTS strategy. Note that in terms of the used forecast error distribution, the factor would be a here, and not $\frac{1}{1+a}$ like before, since the numerically determined pure MTS basestock level ($PureS$) is based on actual demand values (and not forecasted ones). The difference between the two following hypothesis for S^* lies in how they deal with rounding to full batches. The first option always rounds up to full batches, the second option performs a regular arithmetic rounding to full batches.

$$S^* = \lceil E(|FCerr|) \cdot PureS \rceil \qquad (5.4)$$

$$S^* = [E(|FCerr|) \cdot PureS] \qquad (5.5)$$

Hypothesis (5.3, 5.4, 5.5) are now tested by comparing their suggested values for FCT^* and S^* to the numerically determined values in the previous experiment. Table 5.4 lists the results.

Table 5.4 Calculated versus numerically determined values for FCT* and S*

FCT* numerically	FCT* calculated	S* numerically	S* calculated by ceiling function (5.4)	S* calculated by arithmetic rounding (5.5)
0.60	0.63	5	5	5
0.60	0.63	4	5	4
0.60	0.63	4	5	4
0.80	0.83	2	2	1
0.80	0.83	2	2	1
0.80	0.83	2	2	2
0.70	0.71	3	3	2
0.80	0.83	2	2	2
0.70	0.71	4	4	3
0.70	0.71	3	3	3
0.90	0.83	2	2	2
0.80	0.83	3	2	2
0.50	0.63	3	5	5
0.70	0.71	4	4	3
0.70	0.71	3	3	3
0.70	0.71	4	4	3

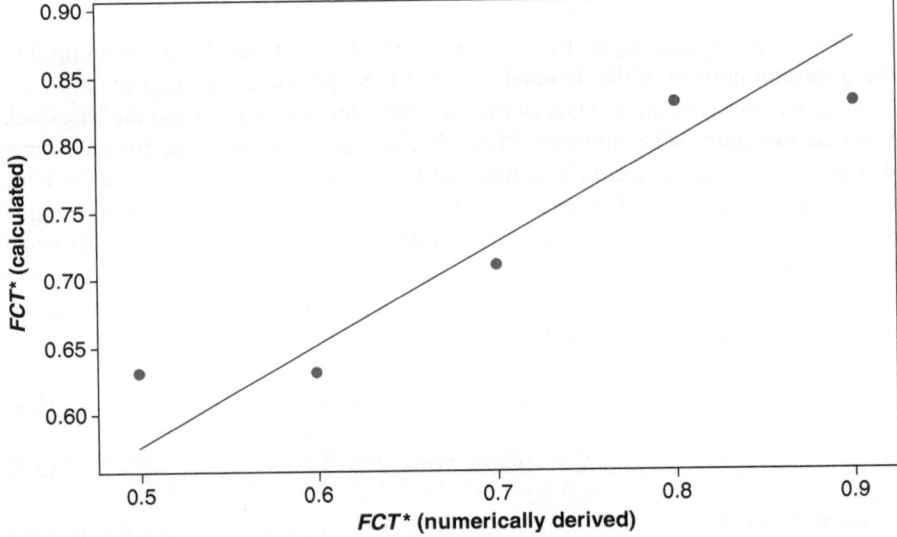

Fig. 5.3 Regression of calculated versus numerically determined values for FCT*

For the numerically versus calculated values of FCT*, a regression analysis has been performed. Its result is displayed in Fig. 5.3. The complete analysis (and all following complete statistical analysis) can be found in Appendix 8.11.3.

For FCT*, the data justifies the hypothesis ($R\text{-}Sq = 88.1\%$, p-value $= 0$). The minor deviations can be largely attributed to the inaccuracy caused by numerically determining FCT* with a resolution of 10 levels (or accuracy of 0.1).

For the case of S^*, both hypothesis were tested (see Appendix 8.11.3) and the first one, which always rounds up to full batches, is found to perform better ($R\text{-}Sq = 73.4\%, p = 0.00$ versus $R\text{-}Sq = 58.0\%, p = 0.00$) and is thus accepted. Based on the two accepted hypotheses above, it can be further concluded that

$$S\%^* = \frac{S^*}{PureS} = \frac{[E(|FCerr|) \cdot PureS]}{PureS} \approx E(|FCerr|) \tag{5.6}$$

and $S\%^*$ converges to $E(|FCerr|)$ with increasing $PureS$ (showed by applying L'Hospital's rule)

$$\lim_{PureS \to \infty} S\%^* = E(|FCerr|) \tag{5.7}$$

Proof:

$$\lim_{PureS \to \infty} S\%^* = \lim_{PureS \to \infty} \frac{S^*}{PureS} = \lim_{PureS \to \infty} \frac{[E(|FCerr|) \cdot PureS]}{PureS} \overset{L'Hospital}{=}$$
$$\lim_{PureS \to \infty} \frac{[E(|FCerr|) \cdot PureS]'}{PureS'} = \lim_{PureS \to \infty} \frac{E(|FCerr|)}{1} = E(|FCerr|)$$

For the relation among FCT^* and $S\%^*$ it can then be stated that

$$FCT^* = \frac{1}{1 + E(|FCerr|)} \approx \frac{1}{1 + S\%*} \tag{5.8}$$

Based on a special and hypothetical distribution for the forecast error, we were able to show that a closed-form approach to parameterize the hybrid MTF/MTS system exists, and that the two parameters FCT^* and $S\%^*$ are associated with each other by a simple relation. Further experiments have been conducted that show the extendibility to arbitrary forecast error distributions (see Appendix 8.11.4).

5.2.3   Extension to the Multi-Product Case

Even though our queuing network model, the optimization procedure, and the simulation model allow for multiple products, all previous experiments in this chapter were made by observing a single product only. In the following, we will extend the results to a multi product environment. For this, we will formulate two assumptions, under which we hypothesize that the previous results are applicable with no further modification. We will examine these hypothesis empirically and then discuss the cases in which the assumptions do not hold.

It is proposed that the results of the experiments above are also applicable in a multi-product environment if:

Table 5.5 Results of the multi-product confirmation experiment

		FCT*		S*	
Forecast error	Demand variability	Analytically	Numerically	Analytically	Numerically
High	High	0.77	0.80	1	1
High	Low (none)	0.00	0.00	5	5
Low (none)	High	1.00	1.00	0	0
Low (none)	Low (none)	1.00	1.00	0	0

- No intermediate products are shared among end products.
- If intermediate products are shared, the demands of the affected end products are not negatively correlated.

Following these assumptions ensures that no risk-pooling effect needs to be considered when allocating and sizing S_{ij}.

An experimental setup to confirm the decision rules and the closed-form solutions for the multi-product environment is defined (see Appendix 8.11.5). The results for one product of the multi-product environment are summarized in the following (Table 5.5).

These results are in line with the previously identified decision rules among pure and hybrid strategies. Moreover, they are in line with the closed-form equations for the parameterization of the hybrid MTF/MTS approach.

If the assumptions above do not hold, the solution suggested by the defined decision rules and closed-form parameterization is not necessarily the best solution anymore. Through considering risk-pooling effects, a further performance increase can be achieved. Then, a full optimization and testing of S_{ij} for all products at all stages as well as modeling of the correlations in the demand model is necessary.

5.3 Analysis of the Performance Increase Achievable with the Hybrid Strategy

5.3.1 Drivers for the Relevance of the Decision among Hybrid and Pure Strategies

In the following, it is examined which conditions favor the application of the new hybrid approach. It is the hypothesis that the achievable performance gain by the transition form the pure to the hybrid strategy is driven by a limited set of identifiable factors. Therefore, first, the resulting data from the first experiment (see Appendix 8.11.2) is used in order to determine the drivers that influence the magnitude of the improvement effect achieved by parameterizing the control approach with the optimal values FCT* and S*. The performance increase (response variable) is calculated as the average performance increase of the optimal

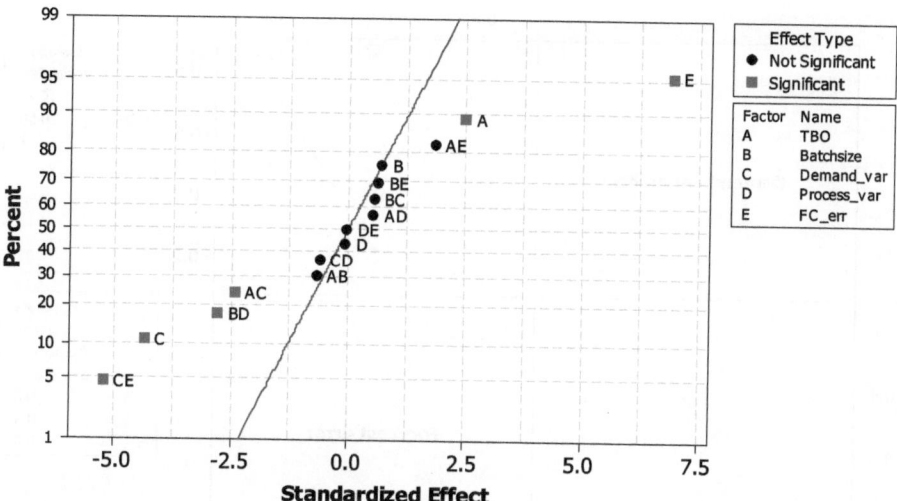

Fig. 5.4 Normal probability plot of standardized effects that influence the average effect of the optimal choice of FCT* and S%* (Alpha = 0.05)

Table 5.6 Drivers for effect size for switching from a pure to the hybrid strategy

Factor	Pure MTS to hybrid effect size (p-value)	Pure MTF to hybrid effect size (p-value)
Demand variability	0.01	0.25
Forecast error	0.20	0.01
Demand variability*Forecast error (interaction effect)	0.82	0.02

strategy compared to 1) pure MTS and 2) pure MTF (last two columns of the result data table in Appendix 8.11.2) (Fig. 5.4).

The graphic shows that the demand variability, the forecast error, and their interaction effect are clearly the most important drivers for the potential improvement magnitude. They will be investigated in more detail below. Moreover, *TBO* has an effect. The longer the time between orders, the more relevant is the right decision among pure MTS, pure MTF, and hybrid MTF/MTS. This is an obvious result, since a longer *TBO* also means higher stocks and thus a larger potential field of influence for the control approach. The batch size and the process variability have no influence on the potential improvement magnitude. Due to their relatively low influence, the two remaining interaction effects ("BD", "AC") are not examined in more detail.

Next, we will investigate the three major drivers, demand variability, forecast error, and their interaction effect, given the scenario, that a hybrid approach is indicated. We therefore use the result data from the second experiment in this chapter (see Appendix 8.11.3). The following table shows the *p*-values for the relevance of each factor for the effect size of going from pure MTS to the hybrid

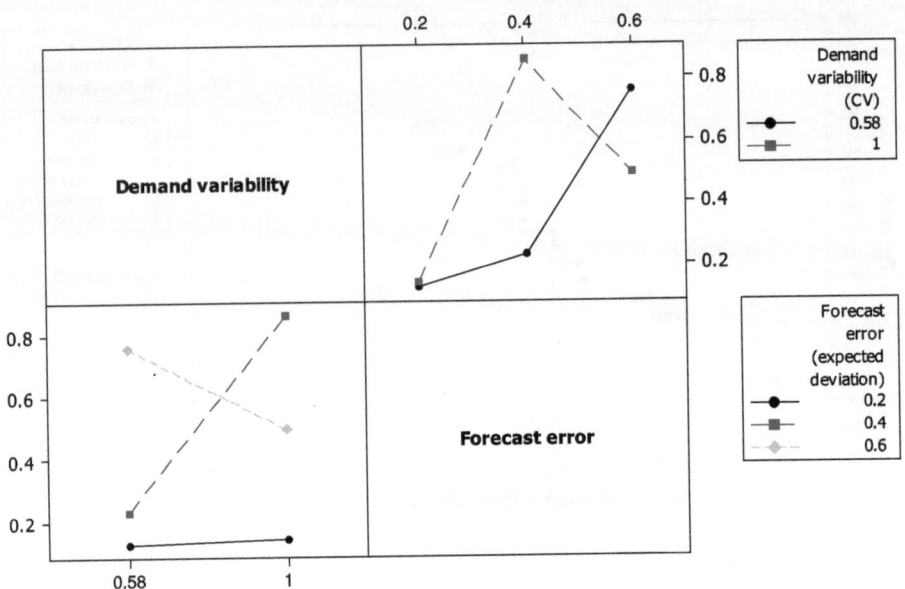

Fig. 5.5 Interaction plot (data means) for the utility improvement from pure MTF to hybrid

approach and for going from pure MTF to the hybrid approach (the detailed statistical analysis can be found in Appendix 8.11.6) (Table 5.6).

It can be seen that (based on confidence level 0.05) demand variability is the only driver for the effect size for the transition from pure MTS to hybrid. For the effect size from pure MTF to hybrid, forecast error and the interaction effect are statistically significant. This result suggests investigating two aspects that are not straightforward in more detail. First, the interaction effect in the MTF to hybrid case and second, which property of the demand distribution exactly causes the differences in effect size.

To better understand the influence of the interaction effect, in the following, interaction plots (Montgomery 2009) are used. To interpret the following interaction plot, the upper right quadrant is used (Fig. 5.5).

It can be seen that with increasing forecast error, also the potential benefit of the hybrid approach tends to grow. However, in the high demand variability case, the effect size drops again from the medium to high forecast error scenario. For the given setting, this can be explained as follows. In the low demand variability case, with growing forecast error, also the benefits that can be achieved by covering for the error with an MTS basestock grow. Thereby, the induced basestock stays, due to the lower demand variability, relatively small. Also in the high demand variability case, introducing an MTS fraction lets the system benefit more with increasing forecast error. However, if the forecast error grows over a certain threshold, the portion of induced MTS-based stock becomes relatively large, due to the high demand variability. This effect then leads to the observed decrease in the potential benefit.

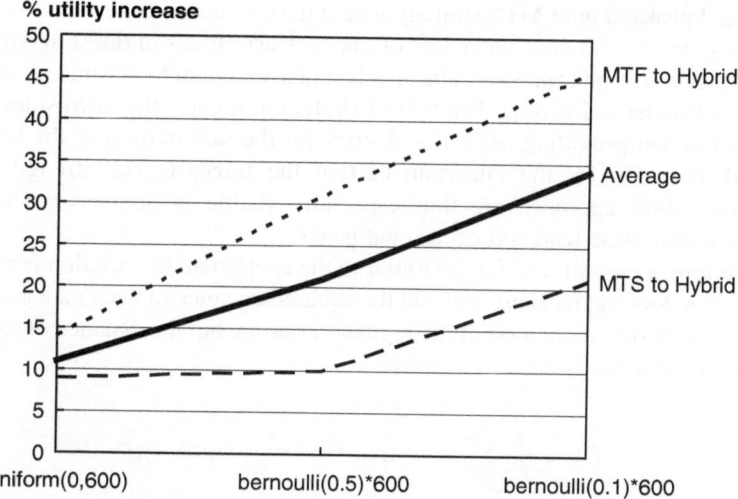

Fig. 5.6 Magnitude of the performance increase of different demand distributions

To better understand the characteristics of the demand distribution that drive the effect sizes, another experiment is set up. Thereby, only the demand distribution is varied on three levels. Two Bernoulli-based distributions and one uniform distribution are set up. The three distributions are chosen such that they exhibit different variances, different kurtosis, and different skewness. The experimental setup and the numerical results can be found in Appendix 8.11.6. Figure 5.6 summarizes the results.

From the graph above, it can be seen that different demand distributions have a significant impact on the potential utility increase achievable with the transition to a hybrid control approach. In one setup, an average performance increase of over 30% is achieved. In the following section, therefore, different statistical measures will be tested for their discriminative power towards the achievable average performance increase. Moreover, based on the gained understanding of the hybrid MTF/MTS approach, a suitable demand variability measure will be proposed.

5.3.2   Derivation of a Demand Variability Measure to Characterize Environments That Favor Hybrid Strategies

In this section, a measure called coefficient of demand variation (CDV) to characterize demand distributions is developed. The measure suggests the magnitude of the potential performance increase achievable by implementing a hybrid PCS.

To define such a measure, first, the root cause for the superiority of the hybrid approach needs to be recapitulated. Assuming a demand environment with occasional peaks, a pure MTS strategy would have to install basestock to cover for the seldom peaks and thus, carry lots of unnecessary stocks in times in which only the demand base load is present. The disadvantage of a pure MTF strategy is the risk of over- or under-delivering. The hybrid strategy mitigates the downsides of each approach by not providing MTS-based stock for the safe portion of the forecasted demand, but only for the uncertain part of the forecast. The strengths of the proposed hybrid approach are thus especially visible in demand environments with a constant base load and occasional peaks.

Therefore, a measure, which is similar to the coefficient of variation is proposed. However, it does not measure and add the squared distance of each data point to the mean (μ), as in the calculation of the regular variation, but the distance of each point to distribution's maximum (*max*) instead.

$$DV = \frac{1}{N} \sum_{i=1}^{N} (x_i - \text{max})^2 = \sigma^2 + (\text{max} - \mu)^2 \qquad (5.9)$$

As noted above, the measure can be expressed and calculated easily by using the first two moments of the demand distribution and its maximum.

Proof:

$$DV = \frac{1}{N} \sum_{i=1}^{N} (x_i - \text{max})^2 \overset{\text{subst: max}=\mu+z}{=} \frac{1}{N} \sum_{i=1}^{N} [x_i - (\mu + z)]^2 = \frac{1}{N} \sum_{i=1}^{N} [(x_i - \mu) - z]^2$$

$$= \frac{1}{N} \sum_{i=1}^{N} (x_i - \mu)^2 - \frac{2}{N} \sum_{i=1}^{N} (x_i - \mu)z + \frac{1}{N} \sum_{i=1}^{N} z^2 = \sigma^2 - \left[\frac{2z}{N} \sum_{i=1}^{N} x_i - \frac{2z}{N} \sum_{i=1}^{N} \mu \right] + z^2$$

$$= \sigma^2 - [2z\mu - 2z\mu] + z^2 \overset{\text{subst: }z=\text{max}-\mu}{=} \sigma^2 + (\text{max} - \mu)^2$$

$$(5.10)$$

The resulting measure (*DV*) can then be standardized analog to the proceeding applied for the coefficient of variation, yielding what will be referred to as coefficient of demand variation (*CDV*)

$$CDV = \frac{\sqrt{DV}}{\text{max}} = \frac{\sqrt{\sigma^2 + (\text{max} - \mu)^2}}{\text{max}} \qquad (5.11)$$

The resulting measure *CDV* shows a high predictive quality. In Fig. 5.7, a scatter plot together with a linear regression between the proposed measure and the average magnitude of the performance increase is displayed. The analysis is based on the experiment and the results described in Appendix 8.11.6. For comparison, other characterizing functions for the demand distribution are shown (coefficient of variation, standard deviation, skewness, and kurtosis) and a regression analysis is performed.

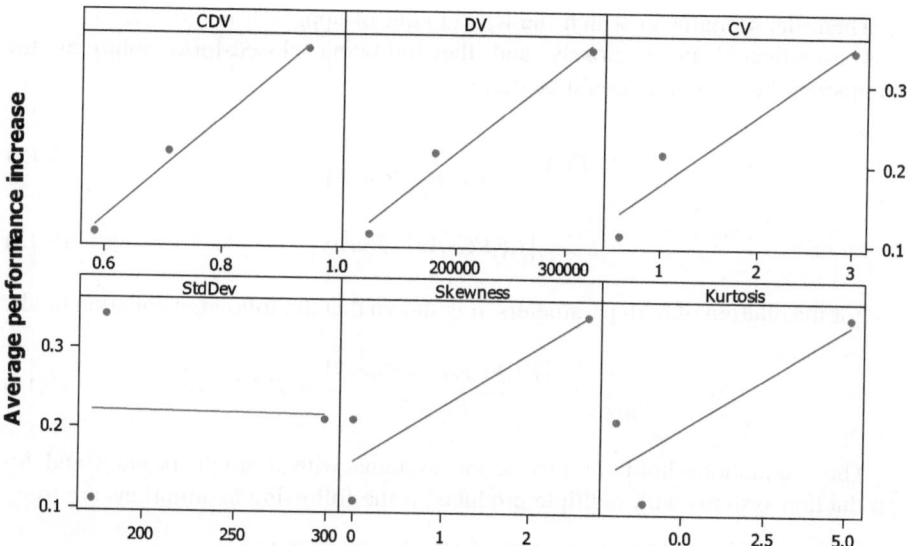

Fig. 5.7 Scatter plot of the proposed demand variability measure along with other distribution characterizing functions

The proposed measure provides the best approximation for the average performance increase (R-Sq = 99.1%, for details see Appendix 8.11.6). In this setting, a *CDV* close to 1, which characterizes a demand scenario with an order peak only in every tenth forecasting period, would enable an average improvement potential larger than 30%. The demand scenario with a uniform demand distribution exhibits a *CDV* around 0.5 and enables an average improvement potential around 10%.

5.4   Summary of Insights

Within this chapter, a series of fractional-factorial-, full-factorial-, and confirmation experiments in order to better understand the physics behind the choice and parameterization of the upstream control approach were performed.

First, the drivers behind the choice between MTS, MTF, and the hybrid MTF/MTS approach were analyzed and demand variability, forecast error, and their interaction effect identified as driving factors behind the decision. The following three decision rules could be identified:

Rule 1: If there is no demand variability, a pure MTS strategy is indicated.
Rule 2: If there is demand variability but no forecast error, a pure MTF strategy is indicated.
Rule 3: If demand variability and forecast error are present, a hybrid MTF/MTS strategy is indicated.

Then, the scenario in which the hybrid control approach is indicated (Rule 3) was investigated more closely and the following closed-form solutions for parameters FCT* and S* could be derived.

$$FCT^* = \frac{1}{1 + E(|FCerr|)} \tag{5.12}$$

$$S^* = \lceil E(|FCerr|) \cdot PureS \rceil \tag{5.13}$$

For the relation of both parameters, it is shown that the following equation holds.

$$S\%^* = \frac{S^*}{PureS} = \frac{\lceil E(|FCerr|) \cdot PureS \rceil}{PureS} \approx E(|FCerr|) \tag{5.14}$$

These equations hold for production systems with a single product and for production systems with multiple products, if the following assumptions are true:

- No intermediate products are shared among end products
- If intermediate products are shared, the demands of the affected end products are not correlated.

This significantly simplifies the application of the PCS engineering framework in practice since the solution space that needs to be covered by numerical methods within the proposed optimization procedure is reduced.

To be able to quickly assess the practical value of the transition from a pure to a hybrid MTF/MTS approach, the driving factors behind the magnitude of the performance increase of the transition were elicited. The drivers are demand variability, time between orders, and the size of the forecast error. The following demand variability measure was introduced, which captures the characteristics leading to a performance gain by the transition to hybrid control better than other tested statistical measures.

$$DV = \frac{1}{N} \sum_{i=1}^{N} (x_i - \max)^2 = \sigma^2 + (\max - \mu)^2 \tag{5.15}$$

After experimentally enhancing the knowledge on the proposed PCS engineering approach, next, its applicability and the power of the hybrid MTF/MTS system will be demonstrated in a case study from the high-tech industry.

Chapter 6
Case Study from the Electronics Manufacturing Industry

6.1 Case Introduction and Specific Challenges

6.1.1 Introduction to the Business and Manufacturing Process

High-tech companies that offer development, production, logistics, and after-sales services to electronics original equipment manufacturers (OEMs) are referred to as electronics manufacturing services (EMS) providers. Due to their explicit specialization on providing operations services in a highly competitive market, the tolerance for poor operational performance is low. The exigencies posed on the PCS are accordingly high. Moreover, the PCS needs to cope with a high level of structural and dynamic complexity, which leads to major challenges in PCS design. The structural complexity is mainly driven by a high number of different customers, products, and processes. The dynamic complexity is caused by short customer lead times, inaccurate forecasts, high demand variability, and hard to control production processes.

The EMS company used for illustration here operates plants in Europe and Asia. Out of each plant, they serve over 150 different customers with over 500 different end-products. The products are inherently complex since they consist of a large number of components and sub-assemblies. Some components and sub-assemblies are shared among products; others are dedicated. The production system is characterized by a complex convergent material flow. The basic processes and the main and numerous alternative material flows are displayed in Fig. 6.1. The first four steps are based on the surface mount technology (SMT) (Prasad 1997). After gathering all necessary components for a batch of PCB boards ("Kitting"), the components are mounted by an SMT-line either single-sided (A-side only) or both-sided (A- and B-side). Besides the vision inspection that is integrated in the SMT line, an additional X-ray-based check ("SMT Test") is performed on PCBs that host components connected by a ball grid array (BGA). Next, the less machine but more manual work driven steps follow. Components that are not suitable for SMT

C. Karrer, *Engineering Production Control Strategies*, Management for Professionals, DOI 10.1007/978-3-642-24142-0_6, © Springer-Verlag Berlin Heidelberg 2012

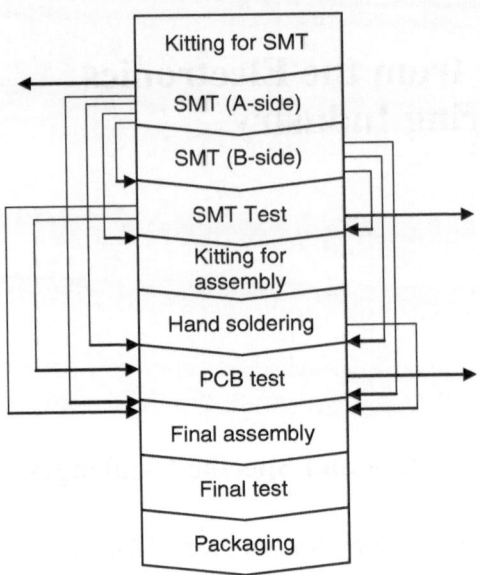

Fig. 6.1 Process flow overview

mounting (e.g. due to their shape or weight) are kitted and hand soldered in U-shaped cells. Afterwards, a function test is performed on the PCB before it goes into the final assembly. There, one or more PCBs are connected and deployed in their housing. After a final test (e.g. burn-in testing), the products are packaged and shipped.

Note that most processes in the observed production system have changeover times between different product variants and thus require batch production. Moreover, the cycle times of a process can vary largely with the type of product. Both aspects further drive the complexity of PCS engineering in this case.

6.1.2 Identification of the Improvement Need and Specific Challenges

Historically, production control was completely based on a standard MRP system. However, the system was not able to cope with the previously described complexity, leading to increased WIP, improvable delivery performance, and problems to keep up high productivity due to frequent re-planning and missing material. The former PCS design was driven by the decision for a standard ERP system and not the result of a proper PCS engineering. The company therefore decided to take a rigorous analytical approach and to engineer a PCS to its needs. Also in this case, the PCS question can be broken down to the three key questions our PCS engineering framework focuses on:

- Limitation of WIP between process steps, aggregation of processes into planning segments (setting K_i)
- Positioning of the order penetration point (setting OPP_j)
- Design of the OPP upstream control (setting FCT_j and S_{ij})

One value stream that covers roughly 20% of the total revenue volume of the plant has been nominated as pilot and will be considered in the following. In the course of a Lean transformation that was executed before this case study has been started, the first two questions were approached in detail. Planning segments were merged into cells with physically co-located workplaces. WIP limits were introduced and implemented visually as FIFO-lanes on the shopfloor in front of each planning segment. The OPP has been positioned after SMT Test. Being not aware of the option to also use MTF or even a hybrid MTF/MTS system together with limited buffers and a defined OPP at that time, for upstream control, a Kanban and supermarket-based pure MTS system has been implemented, strictly following the recommendation of standard Lean thinking. The parameterization of the developed design was not the result of the application of a PCS engineering framework, but the result of a pragmatic, however costly, trial and error approach. For example, to determine K_i, it was started with a high number of K_i that was then stepwise reduced until the system capacity has not been sufficient anymore. The impact of this first PCS engineering cycle on operational performance has been significant.

In this first approach, the delivery performance could be improved dramatically to a stable value close to 100%. At the same time, work-in-process (WIP) could be reduced by almost 70%, leading to a large amount of freed-up invested capital and shortened production lead times, with follow-on benefits such as shorter quality feedback loops and increased flexibility. Note that the WIP reduction was also achieved by other efforts like for instance the disposal of intermediate products waiting for being reworked.

However, in this configuration, several issues remained, which will constitute the focus of the following work. Due to the high number of product variants with peaked demand, the SMT supermarket in front of the OPP, induced by the pure MTS strategy, reached an enormous size of roughly 5,000 PCBs, only for the observed value stream. It consumes valuable space in production and leads to a high capital expenditure. Obviously, after the already good results achieved in the area of the first two key questions, for the management, the biggest improvement need lies in upstream control. The following case study therefore has its focus on the third key question (upstream control approach) and the SMT related processes. Moreover, the case study will be used to validate the recommendations of the PCS engineering framework regarding levels of K_i against the empirically chosen and proven suitable values. The same validation is performed regarding the basestock levels of a pure MTS system.

6.2   Model Development and Parameterization

6.2.1   Production System Model

In the following, the development of the simulation model is described. Thereby, the focus lies on the high-level model and the intricacies of the specific case. A detailed documentation of all model parameters cannot be provided due to confidentiality agreements with the company the case study is based on.

The processes in the observed value stream were aggregated, according to the earlier defined rules (see Sect. 3.1.2), into planning segments as displayed in Fig. 6.2.

In the investigated SMT section of the production, 16 intermediate products (PCBs) are produced and assembled into 9 different end-products according to the matrix given in Fig. 6.3. Moreover, they possibly pass up to four of the above defined planning segments. The allocation of the PCBs to the planning segments is

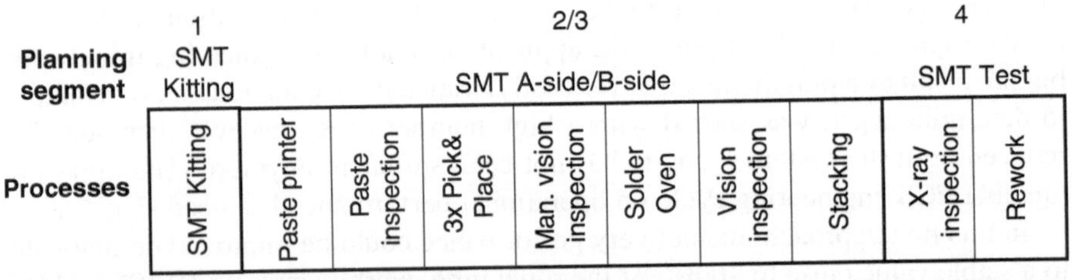

Fig. 6.2 Aggregation of processes into planning segments

PCB	1	2	3	4	5	6	7	8	9	PS 1	2	3	4
1							x			x	x	x	
2				x						x	x	x	x
3	x		x			x				x	x	x	x
4								x		x	x		
5						x				x	x	x	x
6						x				x	x	x	
7						x				x	x	x	
8						x				x	x		
9								x		x	x		
10	x									x	x		
11	x									x	x	x	
12			x			x				x	x	x	
13		x								x	x	x	x
14					x					x	x	x	
15		x								x	x		
16						x				x	x		

Fig. 6.3 Allocation of PCBs to end-products (x denotes goes-into relationship)

also shown in Fig. 6.3. It can be concluded that some end-products share PCBs. Thus, later in this chapter, it has to be evaluated whether a (negative) demand correlation exists and impacts the application of closed-form parameterization rules (see Sect. 5.2.3).

The structure represented by the matrix in Fig. 6.3 has been implemented using the developed simulation framework. The main view of the simulation model which contains the planning segments and the demand model, called 'root object' in the developed simulation framework, can be found in Appendix 8.12.

6.2.2   Parameterization of Planning Segments

When parameterizing the planning segments, several approaches to obtain valid probability distributions were used in this case study. The preferable option was the utilization of data automatically logged by the control units of processes, which can be used to fit valid probability distributions. However, not for each OEE loss category such data was available for the time to repair and the time between failures. Then, operators were instructed to record the relevant data over a suitable period (in most cases 6 weeks). If a planning segment contains several processes, it is important that the elicited data represents the performance of the planning segment as a whole and not the performance of single processes. This can be achieved by measuring at the bottleneck (if clearly identifiable) or at the end of the line. In some cases, it was not possible to obtain data for the time-between-failures. Then, the exponential distribution was assumed and parameterized over the availability equation as proposed in (3.16).

The statistical modeling of kitting operations turned out to be challenging. Therefore, the chosen approach will be explained in more detail. The timing and analysis of single picks for a kit within the observed SMT component warehouse showed that the exponential distribution provides a statistically valid approximation for the time to perform the picking of one component type. Since the largest effort is actually finding the component and allocating the cart containing the kit next to it, the number of component types per kit is the driver for the time needed to complete a kit, and not the total number of components picked. For the transition from modeling picks of one type to modeling the time for a whole kit, a helpful relation among the exponential and the gamma distribution can be exploited. If random variable $X \sim exponential(a)$ describes the picking time for one component type, the needed time for a full kit with C component types can be described as

$$\sum_{i=1}^{C} x_i \sim gamma(C, a) \tag{6.1}$$

in case x_i are independent (Bosch 1998), which can be assumed for picking times. The SMT Kitting planning segment has then been parameterized using this idea.

In this production system, SMT operates continuously in three-shift mode whereas SMT Kitting and SMT Test operate in two-shift mode with no night shift. The transport batch sizes range for most products between 24 and 48. The simulated OEE levels were successfully validated against OEE levels observed in practice.

6.2.3 Parameterization of the Demand Model

In this case, *TBFC* was 1 month and the *TBO* ranged between 1 and 2 weeks depending on the product variant. The forecast that is received for the next month is delivered 3 weeks before it becomes valid. This implies that the maximum possible *FCadvance* is 3 weeks. Since only the OPP upstream part is considered, *CLT* is set to its minimal value, 1 (min).

The demand and forecast data available is naturally based on end-products. Thus, to obtain demands for the observed part of the value chain, a BOM resolution has to be performed. Note that PCBs 3 and 12 go into several end products whose demands have to be aggregated. The distribution fitting for the demand data has been performed using data of the past 6 months. In some cases, no theoretical distribution provided a valid fit. In these cases, empirical probability distributions[1] were implemented.

Since we found intermediate products in our value stream that are shared by end-products, it is indicated to analyze, whether a correlation among the demands exists. However, in the investigated part of the process chain (SMT part), for each intermediate product, already the aggregated demand over all end products is considered (due to the BOM resolution). Thus, possible correlation has no impact on upstream design. For illustrative purposes, we will proceed with the correlation analysis anyhow.

Table 6.1 shows the Pearson correlation coefficient[2] (Bosch 1998) for the end product demands and the *p*-values indicating whether there is a statistically significant correlation or not.

Based on a significance level of 0.05, it can be concluded that only one (negative) correlation exists, namely among end products 5 and 6. However, these two products do not share any PCBs (see Fig. 6.3) and thus, their correlation will not influence the optimality of our decision rules (see Sect. 5.2.3).

To model forecast error, we used a relative forecast error measure as defined in Sect. 3.4.1 and removed the bias from the observations. Then, we fitted theoretical, or in some cases, empirical distributions.

---

[1] Instructions to define empirical probability distributions can be found in Law and Kelton (2008)

[2] $\dfrac{Cov(X, Y)}{StdDev(X) \times StdDev(Y)}$

Table 6.1 Correlation analysis of end-product demands

End product	1	2	3	4	5
2	0.148				
	0.813				
3	0.138	−0.013			
	0.794	0.983			
4	0.019	−0.738	0.234		
	0.971	0.155	0.614		
5	−0.579	−0.017	−0.644	−0.480	
	0.228	0.978	0.119	0.275	
6	0.169	−0.395	0.604	0.561	−0.854
	0.749	0.511	0.151	0.190	0.014

Cell content: Pearson correlation p-value

The built simulation model will be further validated during the PCS design in the following section by comparing its suggestions to the by trial-and-error optimized part of the PCS already in place.

6.3 PCS Design

6.3.1 Strategy Derivation

In the following, the simulation model and the closed-form solutions derived in Sect. 5 are applied to formulate a PCS for the case study. First, the simulation model will be used to derive optimal values for K_i, which are compared to the ones applied in practice. Then a policy for upstream control is formulated and the impact is evaluated.

To run the optimization of K_i according to equation (4.6), the peak and the expected productivity within the shortest order period TBO_j need to be determined. This is done using the order history of the past 6 months, which has also been used in fitting the demand model. The demand values are multiplied with the work content of the observed process steps and divided by the available time, as described in (4.5). From this, $prod^{DemandedAvg}$ can be calculated as average, and $prod_{TBO}^{PeakDemanded}$ as observed maximum. The bounds of K_i are evaluated to $k_i^{upper} = 20$, $k_i^{lower} = 0$ $i = 1, .., 3$. The p-value of the K_i optimization is set to 0.99. Running the optimization[3] leads to a K_i of 10 between SMT Kitting and SMT and to a K_i of 8 between SMT and SMT Test. The amount between SMT Kitting and the SMT line (10 batches) can be transformed into its time supply equivalent by dividing it by the expected demand rate. Then, the 10 batches equal

---

[3] Simulation runtime 1,200,000 (min), 3 replications

roughly 25 h supply, which is very close to the goal of 1 day that has been defined by trial-and-error and been tested successfully during the Lean transformation. The same is true for the buffer between SMT and SMT Test.

Next, FCT_j and S_j are derived using the closed-form solutions from the previous chapter. In this case, both demand variability and a forecast error are present. According to the decision rules we developed in Sect. 5.1.2, a hybrid MTF/MTS approach is indicated. Therefore, the coefficient of demand variation (CDV) is calculated according to (5.11) in order to get an estimate for the expected impact size of the transition from pure MTS to the hybrid control (Fig. 6.4).

The CDV reaches values around 0.5, which suggests that the demand variability is sufficient to enable a moderate but relevant performance increase by applying the hybrid MTF/MTS approach. Thus, it is continued and the policy parameters are calculated.

For all fitted forecast error distributions, $q_{0.5} = 0$ holds and thus, the application of the equations for the closed-form determination of FCT^* and S^*, as described in (5.3, 5.4), is valid. Table 6.2 summarizes the resulting optimal system configuration, also including the optimal basestock levels of a pure MTS strategy (*PureS*). *PureS* were determined experimentally by simulation. Instead of simulation, also the basestock levels of the currently installed pure MTS system could be used. The optimization has been performed anyhow in order to further validate its results against the *PureS* values of the MTS system in place and to obtain a simulation model to estimate the impact of the hybrid control approach later. Comparing the basestock levels of the pure MTS strategy yielded by simulation to those currently used in practice, which were determined by a trial and error approach (remove basestock until delivery is not secured anymore) led to the following result. The average deviation over all product variants is found to be <3% which again, like with the K_i values, indicates that the PCS engineering framework operates accurately.

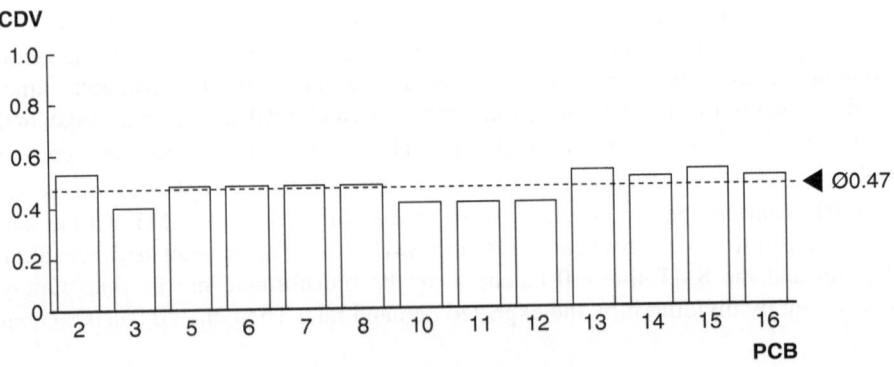

Fig. 6.4 Coefficient of demand variation (CDV) for each variant

Table 6.2 Summary of upstream control parameters

PCB	E(/FCerr/)	FCT*	PureS	S*
1	N/A	0.00	0	0
2	0.24	0.81	6	2
3	0.42	0.70	19	8
4	N/A	0.00	0	0
5	0.50	0.67	8	4
6	0.50	0.67	11	6
7	0.50	0.67	10	5
8	0.50	0.67	11	6
9	N/A	0.00	3	3
10	0.16	0.86	3	1
11	0.16	0.86	4	1
12	0.30	0.77	9	3
13	0.15	0.87	9	2
14	0.50	0.67	8	4
15	0.15	0.87	9	2
16	0.50	0.67	7	4

For the end products which use PCB 9, no forecast is provided. Therefore, FCT^* is set to 0 and they have to be processed in a pure MTS system, what implies $S^* = PureS$. PCBs 1 and 4 are not active.

Next, the potential impact of the implementation of a hybrid strategy as proposed above on the production system is evaluated by simulation.

6.3.2 Impact Evaluation

The pure MTS system configuration and the hybrid MTF/MTS system configuration are run for 1,200,000 (minutes) each and the results are compared in order to evaluate the performance improvements achievable by introducing the hybrid MTF/MTS approach. The delivery performance (alpha service level) is in both cases sufficiently high with regard to the value function and is optimized to a value over 95%. The inventory is measured as average level of the supermarket in front of the OPP.[4] The subsequent table summarizes inventory quantities and values for both approaches and all PCBs as well as the impact in pieces and monetary value. As per piece value, the accounting value for each variant after SMT Test is used (Table 6.3).

---

[4] Note that this way of measuring inventory differs from the approach taken during the definition of CDV in Sect. 5.3.2. There, total WIP in the system was considered, which leads to a lower relative impact of the transition to hybrid control

Table 6.3 Impact summary of pure MTS versus hybrid MTF/MTS

PCB	Per piece value [EUR]	Inventory pure MTS [Pieces]	Value pure MTS [EUR]	Inventory Hybrid [Pieces]	Value Hybrid [EUR]	Impact (Delta) [Pieces]	Impact (Delta) [EUR]
1	N/A	0	0	0	0	0	0
2	671	123	82.495	75	50.302	48	32.193
3	456	360	164.037	264	120.294	96	43.743
4	N/A	0	0	0	0	0	0
5	203	350	71.169	258	52.462	92	18.707
6	109	444	48.218	333	36.163	111	12.054
7	20	378	7.376	319	6.225	59	1.151
8	12	419	5.237	338	4.224	81	1.012
9	N/A	1,337	0	1,337	0	0	0
10	11	110	1.231	70	784	40	448
11	32	155	4.950	64	2.044	91	2.906
12	33	361	11.833	246	8.064	115	3.770
13	343	289	99.214	127	43.599	162	55.615
14	148	299	44.312	251	37.199	48	7.114
15	12	303	3.594	147	1.744	156	1.850
16	11	251	2.762	181	1.991	70	770
Sum		**5,179**	**546.427**	**4,010**	**365.093**	**1,169**	**181.334**

Naturally, for variant 9, which has to be processed according to a pure MTS since no forecast data is available, the impact is zero. The improvement potential is summarized graphically in Fig. 6.5.

The hybrid MTF/MTS strategy outperforms the currently applied pure MTS strategy and would enable an inventory reduction of 1,169 PCBs or 23%. Translated into monetary value, this effect is even bigger, since apparently, the impact is stronger than average on the expensive PCBs. The value of the inventory could be reduced by 33%, which would equal a capital expenditure (CAPEX) reduction of roughly EUR 180 thousand. This effect quantifies only the book value of the inventory. Further positive effects like the freed-up production space or reduced

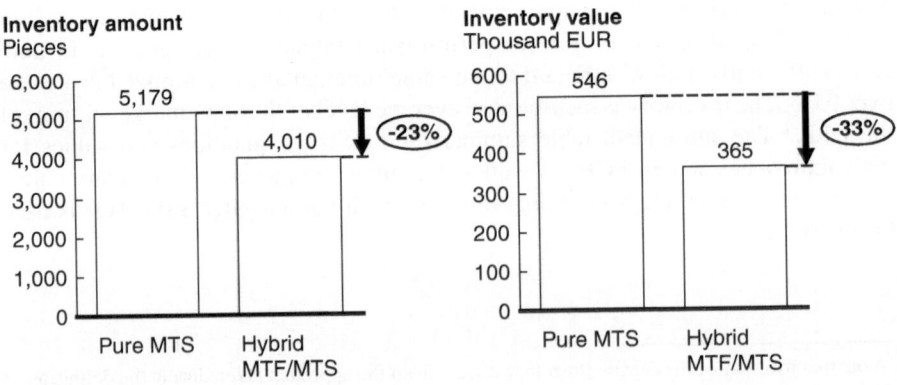

Fig. 6.5 Summary of improvement potential

Cumulated inventory size reduction potential

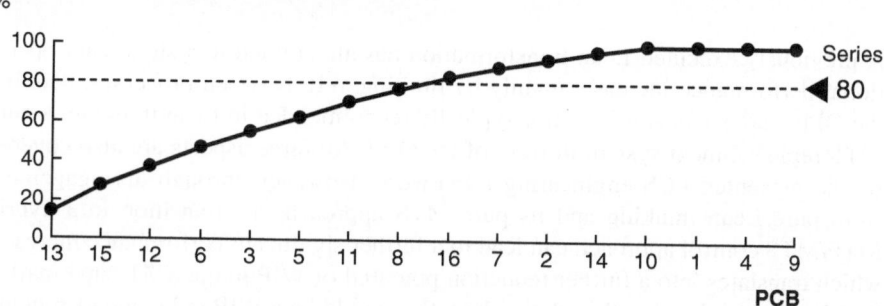

Cumulated CAPEX reduction potential

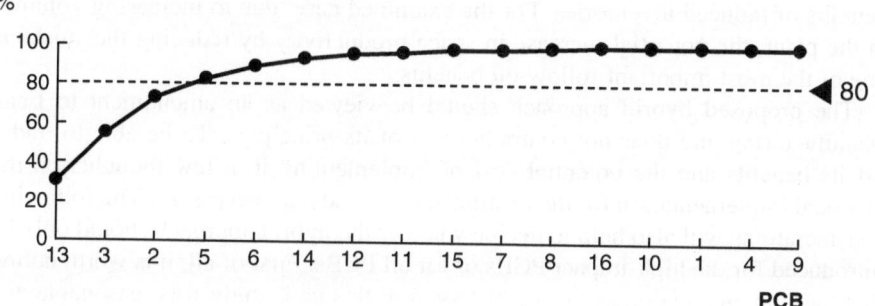

Fig. 6.6 Pareto chart of improvement potential

handling efforts are not quantified. However, these effects can easily reach the same magnitude or even more as the pure inventory value reduction. Especially in the context of the company in the case study, the value of production space is high due to steadily growing production volumes in the examined plant.

Looking at the distribution of the potential over the different PCBs (Pareto graph in Fig. 6.6), it can be seen that 80% of the inventory reduction results from 9 out of 16 PCBs. For the CAPEX reduction, 4 out of 16 PCBs deliver 80% of the potential. For the implementation, which will be discussed in the following, it is thus worth considering whether the hybrid approach should be introduced for all PCBs or only the ones delivering high impact.

The identified potential for the examined value stream can be interpolated to the whole plant. Assuming that the PCS design drivers are similar in the other value streams and that the observed value stream represents around 20% of the total volume, an overall improvement potential of EUR 900 thousand only considering the book value of the PCBs can be hypothesized.[5]

_____

[5] Same improvement opportunity (EUR 181,334 in 20% of plant volume) assumed for remaining plant, leading to EUR 181,334 x 5 = EUR 906 671

6.4 Implications and Implementation Hints

A previously executed Lean transformation has already led to a significant operational performance increase, mainly by limiting WIP between processes, defining the OPP, and establishing, as it is typically recommended in Lean thinking, a pure MTS replenishment system in front of the OPP. All three aspects are also covered by the presented PCS engineering framework. However, through disengagement from pure Lean thinking and its pure MTS approach, the transition to a hybrid MTF/MTS control approach can lead to a further significant performance increase, which translates into a further reduction potential of WIP in the SMT supermarket by 33%. Interpolated to the whole plant, this would be a WIP reduction of roughly EUR 900 thousand, only considering the book value of the stock and no follow-on benefits of reduced inventories. For the examined case, due to increasing volumes in the plant, the potential increase in space productivity by reducing the stocks is one of the most important follow-on benefits.

The proposed hybrid approach should be viewed as an amendment to Lean Manufacturing and does not contradict any of its principles. To be able to trade-off its benefits and the potential cost of implementing it, a few thoughts on the practical implementation for the examined case study are necessary. The following considerations will also help to decide whether the hybrid approach should only be introduced for the high-impact PCBs or for all PCBs. First of all, it is worth noting that the implementation sequence observed in this case study was reasonable and should also be applied in other settings. It makes sense to start with a pure MTS system in front of the OPP until production is stabilized. Then, the more sophisticated hybrid approach, which can build on the basic infrastructure of the MTS system, is introduced. In the course of its implementation, the basestock levels are reduced from $PureS$ to S^*. At the same time, the MTF-based orders need to be generated and represented in production. One approach to represent them could be the introduction of different types of Kanban cards as described in Fig. 6.7.

Besides the regular MTS-based Kanban cards, MTF-based cards are introduced, which are, unlike the MTS Kanban cards, discarded whenever a batch of PCBs is removed from the supermarket. They are generated on the arrival of a new forecast and then, pre-drawn by $FCadvance$, leveled over the TBF, and released in production. Thus, three types of production instructions exist: customer order-based (PI 1), MTS-based (PI 2), and MTF-based (PI 3). Their release and generation in accordance with the arithmetic described before are performed by the production planning and control authority (in the company of the case study called "order coordinator").

Further tasks of the production control authority are the continuous adaption of basestock levels S^*, the adaption of the buffers K_i and, if indicated, the reallocation of the OPP. All these tasks actually require the production planning and control authority to continuously monitor the PCS design drivers and to execute the proposed PCS engineering process or parts of it as changes occur or are expected to occur.

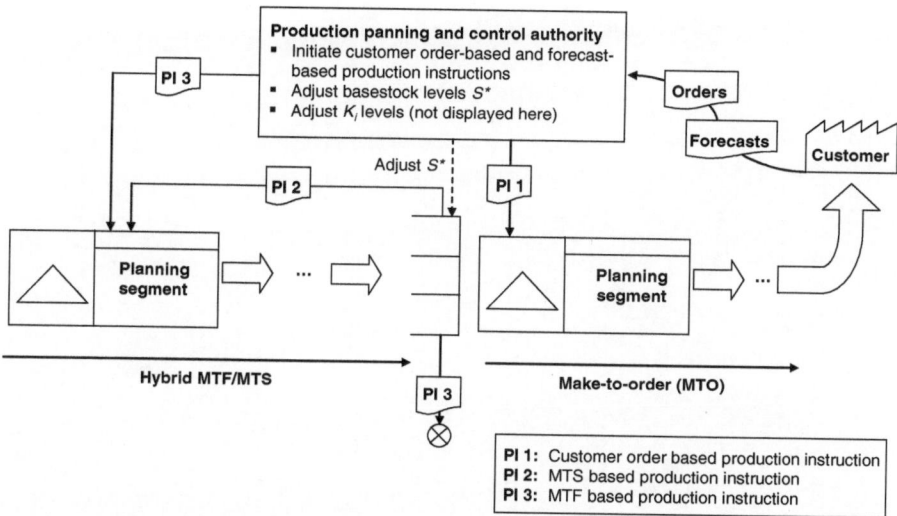

Fig. 6.7 Implementation example of information flow

From this consideration it is obvious, that for the implementation of the approach, the most significant investment is training of the production planning and control employees in PCS engineering. Other significant investments are not required. Even though we saw earlier that the major part of the impact stems from a limited number of variants, for the ease of implementation and operation of the system, it is recommended to transform all variants to the hybrid approach. This leads to a homogenous system that is less complex and easier to learn and to handle than a system that incorporates both, a pure MTS and a hybrid MTF/MTS approach.

Comparing the potential benefits of the proposed hybrid control to the implementation requirements, it can be concluded that if the organizational capabilities are sufficient, which is the case in the company under investigation, it is worth implementing the hybrid control approach.

Chapter 7
Conclusion and Further Research

7.1 Research Summary

Many producing companies face raising expectations with regard to their operations performance. At the same time, they have to deal with an increasing complexity of products and processes. Hence, in order to stay competitive or to outperform competition by operations excellence, a proper PCS is indispensible. Within this research, a PCS engineering framework has been developed to help companies with complex discrete production systems to tailor an effective PCS according to their needs. A systems engineering approach that structures the relevant design drivers and is able to cope with the complexity faced in practice has been taken. The developed generic PCS model is based on queuing network theory and is able to answer the three most prominent questions in PCS engineering:

1. To which amount should WIP be limited to form a buffer between consecutive process steps?
2. Where should the order penetration point (OPP) be positioned in the production flow?
3. How should be dealt with the demand uncertainty at the processes before the OPP? Should a make-to-stock (MTS), make-to-forecast (MTF), or hybrid system be implemented?

Starting from a top-down systems analysis, potential design drivers were identified. They fall in three categories, structure-, productions system variability-, and demand variability-based drivers. All driver categories were integrated into the developed generic queuing network model and methods for its parameterization were presented. The proposed queuing network model approach enhances existing approaches in power and practical applicability. The enhancements include in particular a hybrid MTF/MTS approach to deal with demand uncertainty upstream the OPP, the simple and intuitive structure of the approach, and its ability to deal with complex real-life production systems. To optimize the control parameters of

C. Karrer, *Engineering Production Control Strategies*, Management for Professionals,   117
DOI 10.1007/978-3-642-24142-0_7, © Springer-Verlag Berlin Heidelberg 2012

the generic model and to thus derive a customized PCS, a generic objective function that can be adapted to the preferences of the respective company under investigation is used. The optimization of parameters is performed numerically using exhaustive search. Therefore, an object-oriented discrete-event simulation framework is developed. The two key elements in order to build the models are the planning segment objects and the demand model object. Due to its object orientation, the simulation framework can easily be adapted to any production system and its elements be reused. To be able to cope with the large solution space, especially when modeling complex production systems, complexity reduction techniques are applied. Already in the model building phase, complexity is reduced by separating the production system under investigation into disjoint value streams and into what has been defined as 'planning segments'. A three-step engineering process that allows the sequential optimization of three parameter categories instead of a concurrent optimization of all parameters is introduced. The process significantly reduces the size of the solution space. Further examples of reducing the size of the solution space include experiments for the optimal allocation of basestock in which it is shown that under certain conditions, it is optimal to allocate all basestock directly in front of the OPP. Another effective reduction of the solution space that needs to be treated numerically is achieved by the closed-form solutions developed for the upstream control parameters (FCT^* and $S\%^*$).

For upstream control, the third key-question in PCS engineering, the queuing network model that has been elaborated in the course of this research is not only able to decide between a pure MTF with MRP logic and a pure MTS with pull logic, but is also able to realize a new hybrid MTF/MTS system. This hybrid system significantly outperforms the pure strategies under certain conditions. It is investigated experimentally in more detail. By running structured experiments, decision rules among pure MTS, pure MTF, and the hybrid MTF/MTS are elicited. It is shown that they depend solely on the presence of demand variability and forecast error. Moreover, the already mentioned closed-form determination for the parameters of the hybrid control approach could be developed. The closed form solutions reduce the size of the solution space and thus ease the application of the hybrid approach in practice. It is shown how these closed-form solutions can be applied in environments with multiple products and arbitrary forecast error distributions. To be able to identify environments in which the hybrid approach can have significant impact, the driving factors for the magnitude of the improvement achievable by the transition form pure to hybrid strategies are determined. In the course of this, a new measure for demand variability is proposed that characterizes suitable demand distributions better than existing measures of variability or other characterizing functions of probability distributions.

Within the case study from the electronics manufacturing industry, the applicability of the PCS engineering framework is proven and the model validated. It is demonstrated that even though the examined production system has already undergone a PCS engineering effort in the course of a Lean transformation, in which the first two key questions were addressed, the introduction of the new hybrid MTF/MTS control approach yields another reduction of the necessary inventory in the SMT supermarket by over 30% of its current value. Interpolated to the whole plant, this translates into an improvement potential of roughly EUR 900,000 only considering the book value of the inventory. Further follow-on benefits of reduced inventory levels like freed up production space or reduced material handling effort further increase the improvement potential. The case study is concluded by suggesting a concrete implementation approach of the proposed hybrid MTF/ MTS system that fits into the given technical and organizational context.

7.2 Limitations and Further Research Directions

As we pointed out, the presented PCS engineering framework is able to support companies with complex discrete production systems to improve their operational performance and competitiveness. However, the research conducted so far also leads to areas for further investigation. They will be briefly introduced in the following.

In Sect. 4.3.2, we formulated an optimization problem to determine the buffer size K_i between two planning segments. On the one hand, inventories as defined by the PCS can be seen as hedging strategy to protect customers against variability and to enable achieving the needed productivity. On the other hand, Lean thinking stigmatizes inventories as waste and inhibitor for continuous improvement. One key aspect in Lean Manufacturing is to make failures visible and to generate a sufficiently large need for the organization to root-cause-problem-solve them. A production system that relies on a cushion of inventories is much less driven towards efficiency in every process than a system, in which even small failures are visible as major incidents. Thus, there is a feedback loop from process variability to buffer levels, to the problem solving speed, and back to the process variability. This feedback loop and a time perspective could be integrated in the optimization problem. Especially in industries with highly variable processes and high improvement opportunities, taking into account this trade-off could lead to better buffer sizing decisions. However, the quantification and mathematical modeling of this feedback loop seems challenging.

During the formulation of the demand model, a brief excursion to how the customer could be involved in the design was presented (Sect. 3.4.2). There, also the idea of developing a pricing mechanism depending driven by demand variability was raised. An exploration of this idea in the context of the proposed PCS engineering framework could be a starting point for further research with high practical relevance.

In Sect. 5.2.3, we concluded that the defined decision rules and closed-form solutions for upstream control are only valid in multi-product environments, in which the end-product demands are not negatively correlated or if they are, no intermediate products are shared among correlated end products. Otherwise, risk-pooling effects need to be taken into account when allocating and sizing basestock levels. A closer examination of this scenario could be interesting, driven by the objective to obtain decision rules and closed-form solutions also for the scenario with correlated demands and shared parts.

In the case study in Sect. 6.4, we suggested an approach how the developed MTF/MTS hybrid approach can be implemented on the shopfloor. The presented idea could be refined further and the necessary organization be described in more detail. This should include the relevant business processes and supporting IT systems. An important business process in this context is a process taking care of updating the PCS, a topic that has also been mentioned in the context of the PCS engineering process. Updating triggers and responsibilities need to be defined. Moreover, it could also be an interesting enhancement to enable the incorporation of expert estimates on future changes of design drivers using Bayesian statistics. This would enable a proactive PCS design with the ability to anticipate necessary changes.

Also in the basic structure of the PCS engineering framework as presented in Chap. 3, enhancements are conceivable. First, the excluded lot-sizing problem could be integrated as a further variable influencing the trade-off between WIP and productivity. Moreover, in certain production environments, more sophisticated scheduling or dispatching rules for incoming production orders of a planning segment (queue PO_j) could add value. An exemplary scenario would be a production system, in which the optimally leveled order sequence is different from planning segment to planning segment. For this task, especially agent-based approaches as presented in Sect. 2.2.2.4 could be helpful. Furthermore, the presented approach is designed to be run with one forecast value. However, in environments with long production lead times, it could be the case that between the start of production and its completion, additional, more accurate forecast information becomes available. The development of a proceeding to deal with this additional, more accurate forecast information could be another interesting starting point for further research.

The presented ideas for further research conclude this work. Motivated by the significant improvement potential delivered in the presented case study from electronics manufacturing, it is the hope of the author that the developed PCS engineering framework, and the hybrid MTF/MTS control approach enabled by it, find their way into further industrial application.

Chapter 8
Appendix

8.1 List of Figures

C. Karrer, *Engineering Production Control Strategies*, Management for Professionals, 121
DOI 10.1007/978-3-642-24142-0_8, © Springer-Verlag Berlin Heidelberg 2012

8.2 List of Tables

8.3   List of Abbreviations

ADI	Advance Demand Information
API	Application Programming Interface
APS	Advanced Planning System
AWIP	Average Work in Process
BGA	Ball Grid Array
BOM	Bill of Material
C/O	Change Over
CAPEX	Capital Expenditure
CDV	Coefficient of Demand Variation
CONWIP	Constant Work in Process
CV	Coefficient of Variation
EDD	Earliest-Due-Date (dispatching rule)
EKCS	Extended Kanban Control System
EMS	Electronics Manufacturing Services
ERP	Enterprise Resource Planning
EUR	Euro
EWMA	Exponentially Weighted Moving Average
F/C	Forecast
FIFO	First-In-First-Out (dispatching rule)
GKCS	Generalized Kanban Control System
ID	Identifier
JIT	Just-In-Time
LRPT	Longest-Remaining-Processing-Time (dispatching rule)
MAUT	Multi Attribute Utility Theory
MDP	Markov Decision Process
MRP	Material Requirements Planning
MRP II	Manufacturing Resources Planning
MTF	Make-To-Forecast
MTS	Make-To-Stock
N/A	Not Available
NO	Number
OBS	Observation
OEE	Overall Equipment Effectiveness
OEM	Original Equipment Manufacturers
OPP	Order Penetration Point
PAC	Production Authorization Card (System)
PBF	Parts Between Failures
PCB	Printed Circuit Board
PCS	Production Control Strategy
PDF	Probability Density Function
PI 1	Production Instruction (customer order-based)
PI 2	Production Instruction (MTS-based)
PI 3	Production Instruction (forecast-based)
PO	Production order
PS	Planning Segment

(continued)

PSE	Production Systems Engineering
Repl.	Replication
S&OP	Sales and Operations Planning
SE	Standard Error (of fitted value)
SIPS	Standard in Process Stock
SMT	Surface Mount Technology
SPT	Shortest-Processing-Time (dispatching rule)
SRPT	Shortest-Remaining-Processing-Time (dispatching rule)
STDDEV	Standard Deviation
TBF	Time Between Failures
TPS	Toyota Production System
TTR	Time To Repair
UML	Unified Modeling Language
WIP	Work in Process
■	End of Proof

8.4 List of Notations

In the following, the notations used in the formal development of the PCS engineering framework are listed. Notations that are used solely in dedicated sections of the literature survey to describe other approaches are not included. Also notations that are used specifically within the Java based implementation of the simulation framework and within the applied statistical software package are not listed.

$\lceil x \rceil$	Ceiling function, defined as $\min\{n \in \mathbb{Z} \vert n \geq x\}$
$\vert x \vert$	Absolute value, defined as $\vert x \vert = \begin{cases} x, & x \geq 0 \\ -x, & x < 0 \end{cases}$
	alternatively describes the number of elements in a set
$[x]$	Arithmetic rounding function
\<term\>	Term to be replaced by actual content, used in the description of graphical representations
ACT	Actual cycle time
BS	Batch size
c	Constant, used in different contexts
C	Number of component types in a kit
CD	Conditional disposal in queuing network model
CDV	Coefficient of demand variation
c_i	Capacity constant of internal buffer queue IB_i
CLT	Customer lead time
Cov(X,Y)	Covariance for random variables X and Y, $Cov(X, Y) = E[(X - E(X)) \cdot (Y - E(Y))]$
CV(x)	Coefficient of variation with $CV(X) = \frac{StdDev(X)}{E(X)}$
D, d_t	Demand in pieces, random variable and its realization at time t

(continued)

$DelPerf(OPP_j)$	Numerically determined delivery performance achieved with OPP location OPP_j
$DelPerf_{upperBound}$	Upper bound for delivery performance as defined in value function (Sect. 4.1.2).
DV	Demand variation
$E(X) = \mu$	Expected value or mean with $E(X) = \frac{1}{N} \sum_{i=1}^{N} x_i$
FC, fc_t	Forecast value in pieces, random variable and its realization at time t
$FCadvance$	MRP parameter determining how long before their due date forecast-based POs are launched in production
$FCarrivalAdv$	Maximum possible $FCadvance$, determined by actual forecast arrival time
$FCerr, fcerr_t$	Forecast error, random variable and its realization at time t, superscripts indicate calculation method *relative* or *absolute*, and whether bias has been removed, i.e. *relNoBias*
$FCerror_j$	Random variable representing the forecast error of product j
$FCT*$	Optimal value for parameter forecast trust
FCT_j	Control parameter forecast trust for product j
FG	Queue in queuing network model holding finished goods
i	Index referring to planning segments
$I(x)$	Inventory function modeling the current inventory held in queues or manufacturing processes in the queuing network model
IB_i	Queue within planning segment i acting as internal buffer
IC_i	Queue within planning segment i containing internal clearances
j	Index referring to products
K_i	Control parameter Kanbans between planning segments i-1 and i
k_i^{lower}, k_i^{upper}	Lower and upper bounds for K_i
L	Resolution of decision variables expressed as number of levels
$L(x)$	Loss function
M	Number of products in production system
M_i	Queue within planning segment i containing incoming material
max	Maximum value of a random variable or of a set of realizations
MP_i	Representation of the manufacturing process(es) of planning segment i
N	Number of planning segments in production system or other context depended meaning
$\mathbb{N}$	Set of natural numbers
OEE	Overall equipment effectiveness $\in [0, 1]$
$OEEloss$	Random loss period added to the standard cycle time
OP, op_k	Set of order points in one forecasting period, k-th element of the set
OPP_j	Order penetration point of product j
$OrderSize$	Quantity of one customer order
p	p-value used for stochastic optimization and in statistical tests
$P(X \geq x)$	Probability that random variable X reaches a value larger or equal to x
PC_i	Queue within planning segment i containing production clearances
$PO_j^{<Source>}(t)$	Production orders launched at time t for variant j. Can be initiated by two possible sources, either forecast-based (F/C) or customer order-based (O)
PO_i	Queue within planning segment i containing production orders
POS	Set of positive deviations of demands from forecast due to forecast error

(continued)

$PROD$	Productivity measure
$PROD_{TBO}(K_1, ..., K_N)$	Numerically determined productivity levels achieved within TBO with values $K_1, ..., K_N$
$prod^{DemandedAvg}$	Required productivity level to satisfy demand on average
$prod_{TBO}^{DemandedPeak}$	Required productivity level to satisfy peak demand in relevant time horizon TBO
$PureS$	Optimal basestock level in a pure MTS system
$q_r(X)$	Quantile r of random variable X
$S\%*$	Optimal basestock level expressed as percentage of the basestock level of a pure MTS system
$S*$	Optimal basestock level (subscripts i and j left out)
$s.t.$	Subject to (in the formulation of optimization problems)
SCT, SCT_{ij}	Standard cycle time, for planning segment i and product j
S_{ij}	Basestock in planning segment i for product j
SPT	Standard processing time
$Sreach*$	Time supply of optimal basestock level under assumed peak demand
SSt	Synchronization station in queuing network model
$StdDev(X)=\sigma$	Standard deviation with $StdDev(X) = \sqrt{Var(X)}$
T	Target value used in loss functions, point in time, or other context specific meanings
t	Point in time
$TBFC$	Time between forecasts
$TBO, TBO_{op_k}^{op_{k+1}}$	Time between orders, random variable or constant, time between orders of order points op_k and op_{k+1}
$Utility*$	Utility achieved with optimal configuration ($FCT*$ and $S*$)
$UtilityImprMTF/$ $UtilityImprMTS$	Percentage utility increase achieved by optimal solution compared to pure MTF or MTS
$UtilityMTF/UtilityMTS$	Utility achievable with pure MTF/pure MTS strategy
$V(x_1, ..., x_n)$	Multi attribute value function of attributes $x_1, ..., x_n$
$v_i(x_i)$	Single dimensional value function
$Var(X) = \sigma^2$	Variance with $Var(X) = \frac{1}{N} \sum_{i=1}^{N} (x_i - E(X))^2$
VM_i	Production system variability measure for planning segment i
$VMPS$	Aggregated production system variability measure for the whole production system
$VSSt$	Variant-specific synchronization station in queuing network model
w_i	Weight used in multi attribute value functions
$WIP(K_1, ..., K_N)$	Numerically determined function representing the average WIP resulting from values K_i
$\mathbb{Z}$	Set of integers (negative, zero, positive)

8.5   Complex Production Systems

In general, a system is perceived as complex if it consists of a large number of entities that interact in a complicated, non-intuitive way, what makes it hard to understand or describe. A system's complexity is driven by its static structure and its dynamic behavior. Complex systems exhibit for instance a structure with a high number of different elements and relationships and a dynamic behavior, characterized by stochastic influences and non-linearities (Deshmukh et al. 1998).

In the context of PCS engineering, the following exemplary static and dynamic complexity drivers can be identified (Table 8.1).

Table 8.1 Addressed complexity drivers (Adapted from Scholz-Reiter et al. (2006), Deshmukh et al. (1998), Dobberstein (1998), and Scherer (1998))

Structural	Behavioral/dynamic
• High number of different products or components from which the final product is assembled	• Suppliers reliability (delivery performance and quality)
• High number of customer orders	• Process reliability (breakdowns, scrap rate, variation of processing times, variation of changeover times)
• High number of different process steps Arbitrary bill-of-material (BOM) structure (linear, convergent, divergent, or general product structure)	• Forecast accuracy
	• Agreed and required flexibility by customer (variation in demand, mix, and required lead time)
• Significant variation of work content per process step among different products	
• High variety in routing of material flow (for different products), loops in flow, shared equipment between products or value streams	• Feedback loops, e.g. WIP level and process improvement speed
• Sequence constraint processes (changeover times)	
• Perishable intermediate products	• Internal information reliability (epistemic uncertainty) about status of the production system and its parameters, parameters and environment continuously subject to change, information delay
• Structure of production system not fixed but evolving over time ('agile production systems')	

8.6   Literature Review

Table 8.2

Table 8.2 Overview PCS design literature

Reference	Class	Innovation	Performance evaluation		Major assumptions	Production setting	Benchmarks
			Metrics	Approach			
Beamon and Bermudo (2000)	Hybrid (horizontal)	Push for subassemblies, Pull for final products	Output, lead time, WIP	Discrete-event simulation (SIMAN/ARENA)	FIFO dispatching, no transportation times, 1 piece flow, balanced line but exponentially distributed processing times, demand distributed exponentially	1 product, 5 stages, 3–5 subassembly lines	Push, Pull
Bechte (1984), Bechte (1988)	Hybrid (vertical)	Introduction of workload oriented manufacturing control; similar to Bertrand and Wortmann (1981) but workload of a process determined by directly allocated workload and workload currently in previous processes weighted with a discount factor	N/A	N/A	N/A	N/A	N/A
Bertrand and Wortmann (1981)	Hybrid (vertical)	Introduction of "workload control" – high level MRP system but production orders only started if workload of all necessary processes is below threshold	N/A	N/A	N/A	N/A	N/A
Bonvik and Gershwin (1996)	Pull	Introduction of a CONWIP/Kanban hybrid control policy	Throughput, delivery performance, WIP, backlog	Discrete-event simulation	Deterministic system (besides demand), serial flow, one product, no changeover, one piece flow	Serial 6 station production line with stochastic demand	Kanban, CONWIP
Buzacott (1989)	Pull	Generalized Kanban Control System (GKCS), combination of Kanban and Basestock, predecessor of the PAC (production authorization cards) system (Buzacott and Shanthikumar 1992)	N/A	N/A	N/A	N/A	N/A

(continued)

Table 8.2 (continued)

Reference	Class	Innovation	Performance evaluation				
			Metrics	Approach	Major assumptions	Production setting	Benchmarks
Buzacott and Shanthikumar (1992)	Pull	Introduction of production authorization cards (differentiated Kanban cards) to create advanced different Pull systems, can emulate different Pull systems	N/A	N/A	N/A	N/A	N/A
Cochran and Kaylani (2008)	Hybrid (horizontal)	Junction point allocation problem solved with genetic algorithm	Inventory cost, tardiness cost	Discrete-event simulation	No setup times	Serial production line, multiple products, tube shop	Push, Pull
Dallery and Liberopoulos 2000	Pull	Extended Kanban Control System (EKCS) which combines Kanban and Basestock	N/A	N/A	N/A	N/A	N/A
Gaury et al. (2001), Gaury et al. (2000)	Pull	Proposition of a generic model that allows mapping of Kanban, CONWIP and Basestock and all combinations of them	WIP, delivery performance	Discrete-event simulation (SIMAN)	N/A	Single product, 4 and 8 stage serial production line	CONWIP
Gelbke (2008)	Push	Creation of an agent-based PCS (with focus on order release and repair of inconsistent plans)	N/A	N/A	N/A	N/A	N/A
Glassey and Resende (1988)	Hybrid (vertical)	Starvation avoidance concept as special case of bottleneck scheduling	Throughput	Discrete-event simulation (FabSim)	N/A	Very large scale integrated circuits production, unreliable machines	N/A
Hall (1986)	Hybrid (vertical)	Synchro MRP – classical MRP system combined with 2 card Kanban system	N/A	N/A	N/A	N/A	N/A
Hirakawa (1996)	Hybrid (horizontal)	Push-Pull barrier determined based on forecast error in different time horizons	Variance of inventory level	Simulation	Fully deterministic system	Serial production, one product	Push, Pull
Hodgson and Wang (1991)	Hybrid (horizontal)	Enable each production stage to either Push or Pull	Inventory cost, shortage cost	Analytical by modeling as a Markov decision process	100% raw material availability and quality	3 products, convergent flow, typical iron and steel works process, process breakdowns considered	Push, Pull

(continued)

Table 8.2 (continued)

Reference	Class	Innovation	Performance evaluation Metrics	Approach	Major assumptions	Production setting	Benchmarks
Wang and Hodgson (1992)	Hybrid (horizontal)	Push until parallel production stage merge, Pull afterwards – generalization of Hodgson and Wang (1991)	Inventory cost, shortage cost	Analytical by modeling as a Markov decision process	100% raw material availability and quality	General parallel and/or serial multistage production system	Push, Pull
Huang (2002)	Hybrid (horizontal)	System following the drum-buffer-rope logic; Pull until constraint resource, Push afterwards	Output, bottleneck utilization, average time of WIP in buffer	Discrete-event simulation (Promodel software)	Implicit: constraint resource known and not changing, no transportation or setup time	3 products, 4 types of machines, different but serial flow, shared equipment between products, FIFO dispatching	Push, Pull
Khoo et al. (2001)	Push	Agent-based scheduling system	Make span	Agent-based simulation	N/A	6 product 3 shop scheduling problem, plastic injection moulding	N/A
Li and Liu (2006)	Pull	Integration of lot sizing problem into a WIP minimization problem	WIP, idle probability of bottleneck	Stochastic process	N/A	2 stage production system, multiple products, random processing and changeover times	N/A
Li et al. (2006)	Push	Earliness-tardiness production planning for MRP systems	N/A	N/A	Fully deterministic system	n products, m production stages, T planning periods	N/A
Masin and Prabhu (2009)	Pull	Introduction of AWIP, an optimization approach to design an integrated PCS from Kanban, Basestock, CONWIP and their generalizations	WIP	Discrete-event simulation	Raw material secured, transportation delay for goods and information	1 product, 3 or 6 stations, 200 or 400 products	Kanban, gradient optimization
Mönch (2006), Mönch et al. (2006)	Push	Creation of an agent-based PCS (with focus on scheduling)	Delivery performance, production lead time, throughput	Discrete-event simulation (AutoSched)	N/A	Fictive mid-size semiconductor manufacturing site	FIFO Push
Olhager and Ostlund (1990)	Hybrid (horizontal)	Suggestion and qualitative discussion of locating the Push-Pull junction point according to the customer order point, bottleneck or product structure	N/A	N/A	N/A	N/A	N/A
Pandy and Khokhajaikiat (1996)	Hybrid (horizontal)	Extension of Hodgson and Wang (1991/1992) by uncertain demand and raw material supply	Inventory, raw materials, lost sales, output	Discrete-event simulation (written in Fortran 77)	Process uncertainties (setup times, down times, cycle time) not considered	Serial/parallel four stage hair dryer production line	Push, Pull

(continued)

Table 8.2 (continued)

Reference	Class	Innovation	Performance evaluation		Major assumptions	Production setting	Benchmarks
			Metrics	Approach			
Spearman et al. (1990)	Pull	CONWIP Pull system	WIP, controllability	Discrete-event simulation	Leveled production line (similar processing times for different products/batches)	Serial production system, different products	Push
Suri (1998)	Hybrid (vertical)	Introduction of POLCA control (similar to Synchro MRP). MRP planning combined with POLCA cards (production authorization cads between two processes)	N/A	N/A	N/A	N/A	N/A
Takahashi and Nakamura (2002)	Pull	Reactive Kanban system that adjusts the number of cards based on stable and unstable demand changes detected with control charts	WIP, lead time	Discrete-event simulation	Stochastic processing times, backorder product demand allowed	Linear N stages serial flow, inventories before and after each production stage	Kanban
Takahashi and Soshiroda (1996)	Hybrid (horizontal)	PCS customization – analytical method to derive optimal integration parameter in Push-Pull or Pull-Push hybrid systems (i.e. at which stage to switch the mode)	Output, inventory level	Analytical by solving difference equations	Demand is only uncertainty in the model	Single product, n-stage serial production system	Push-Pull, Pull-Push with different integration parameters
Tardif and Maaseidvaag (2001)	Pull	Reactive Kanban system that adjusts the number of cards based on demand changes	Cost (from WIP)	Queuing model	No uncertainties except demand uncertainty	Single part, single stage	Kanban

8.7 Delimitation against Other Generic PCS

As surveyed in the literature review, three other relevant generic queuing network approaches have been suggested. They include the EKCS (Dallery and Liberopoulos 2000), the GKCS (Buzacott 1989), and the PAC system (Buzacott and Shanthikumar 1992). In the following, the presented queuing network model will be delimited against these approaches by pointing out the key differences.

First, the GKCS, and its enhancement, the EKCS, are pure pull-based systems. They are not integrating the MRP logic or forecast-based production orders. Naturally, they are also not able to represent MTF or hybrid MTF/MTS systems as the suggested generic model. Moreover, they do not explicitly feature the allocation of the OPP, like the presented model. The parameter OPP_j is not explicitly introduced to address this problem. In their original versions, the EKCS and the GKCS were only capable to map single products. Baynat et al. (2002) present an extension to multi-product cases for the EKCS. However, in their suggested approach, the number of necessary queues and synchronization stations scales with the number of products mapped. For complex production systems, this would quickly lead to a highly complex structural model. The presented PCS engineering framework avoids this problem by introducing variant-specific synchronization stations ($VSSt$). Also, the presented approach exhibits a faster information flow. Like the EKCS and unlike the GKCS, demand information (production orders) is directly communicated to each planning segment. In addition, production authorizations are released faster than in the EKCS. The EKCS sends a production authorization to the preceding process only after completing the processing of the job. In contrast, the presented approach immediately releases an authorization (production clearance) in queue PC of the preceding planning segment after the material has been removed for processing and thus avoids the delay of one batch cycle. The resulting problem of uncontrolled WIP within the planning segment is solved with an additional internal control cycle, represented by queues IC. The PCS that can be represented with the EKCS and the GKCS are also included in the proposed approach. However, the integration of these additional aspects comes at the price of a larger and more complex solution space.

Compared to the earlier mentioned powerful PAC approach (Buzacott and Shanthikumar 1992), the presented PCS engineering framework could be easier to deploy in practical application. This hypothesis is based on the more intuitive and simpler structure of the proposed approach, which is oriented at the practical needs of PCS engineering. The number of information items that cycle is lower compared to the variety of tags used in the PAC systems. Their functions are dedicated and the three basic parameter categories can, as we will see later, be optimized sequentially. The proposed approach is more specific (i.e. the PAC system often refers to management of the cell that needs to take decisions) and describes all mechanics involved without compromising on general applicability. Moreover, the PAC approach does not include a parameter like FCT_j and thus does not enable the described hybrid MTF/MTS approach. Even though OPP_j is not explicitly mentioned, also the PAC system could be configured in a way to reveal the OPP.

8.8 Evolution of the Maximum Size Reduction of the Solution Space

1. Initial size of the solution space (Sect. 3.2.4)

$$L^{(N-1)+M+\sum_{j=1}^{M}(OPP_j-1)} \cdot (N+1)^M \tag{8.1}$$

L: Resolution of decision variables expressed as number of levels
N: Number of planning segments
M: Number of product variants
OPP_j: Location of the order penetration point (in front of which planning segment)

2. Stepwise optimization as defined in optimization procedure (Sect. 4.3.1)

$$L^{(N-1)+M+\sum_{j=1}^{M}(OPP_j-1)} \cdot (N+1)^M \Rightarrow L^{(N-1)} + L^{M+\sum_{j=1}^{M}(OPP_j-1)} + (N+1)^M \tag{8.2}$$

3. Separated OPP allocation for each production in step two of the optimization procedure (Sect. 4.3.3)

$$L^{(N-1)} + L^{M+\sum_{j=1}^{M}(OPP_j-1)} + (N+1)^M \Rightarrow L^{(N-1)} + L^{M+\sum_{j=1}^{M}(OPP_j-1)} + M \cdot (N+1) \tag{8.3}$$

4. Basestock allocation only in front of the OPP (Sect. 4.3.4)

$$L^{(N-1)} + L^{M+\sum_{j=1}^{M}(OPP_j-1)} + M \cdot (N+1) \Rightarrow L^{(N-1)} + L^{2M} + M \cdot (N+1) \tag{8.4}$$

5. Separated optimization or upstream control for each product (Sect. 4.3.4)

$$L^{(N-1)} + L^{2M} + M \cdot (N+1) \Rightarrow L^{(N-1)} + M \cdot (L^2 + N + 1) \tag{8.5}$$

6. Closed-form determination of FCT_j and S_{ij} (Sect. 5.2.1)

$$L^{(N-1)} + M \cdot (L^2 + N + 1) \Rightarrow L^{(N-1)} + M \cdot (L + N + 1) \tag{8.6}$$

8.9 Simulation Model Implementation

8.9.1 Planning Segment Graphical Representation

Figure 8.1

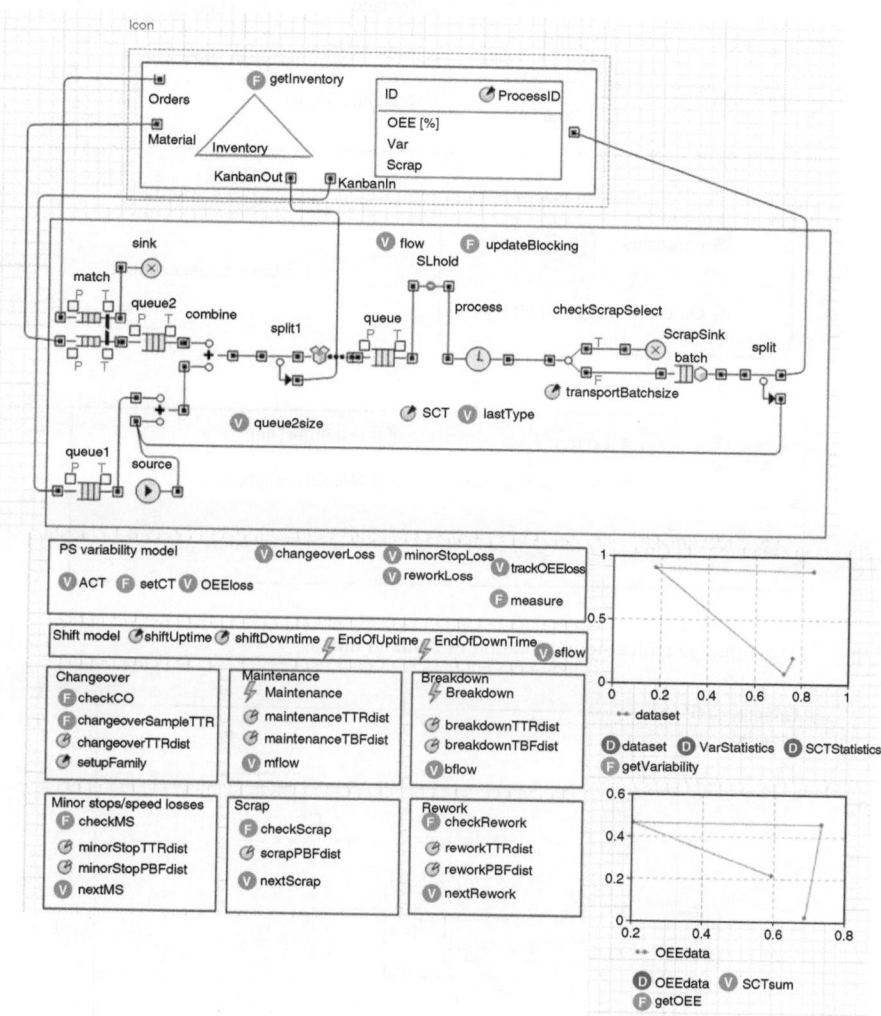

Fig. 8.1 AnyLogic graphical representation of the planning segment model

8.9.2 Demand Model: Graphical Representation

Figures 8.2 and 8.3

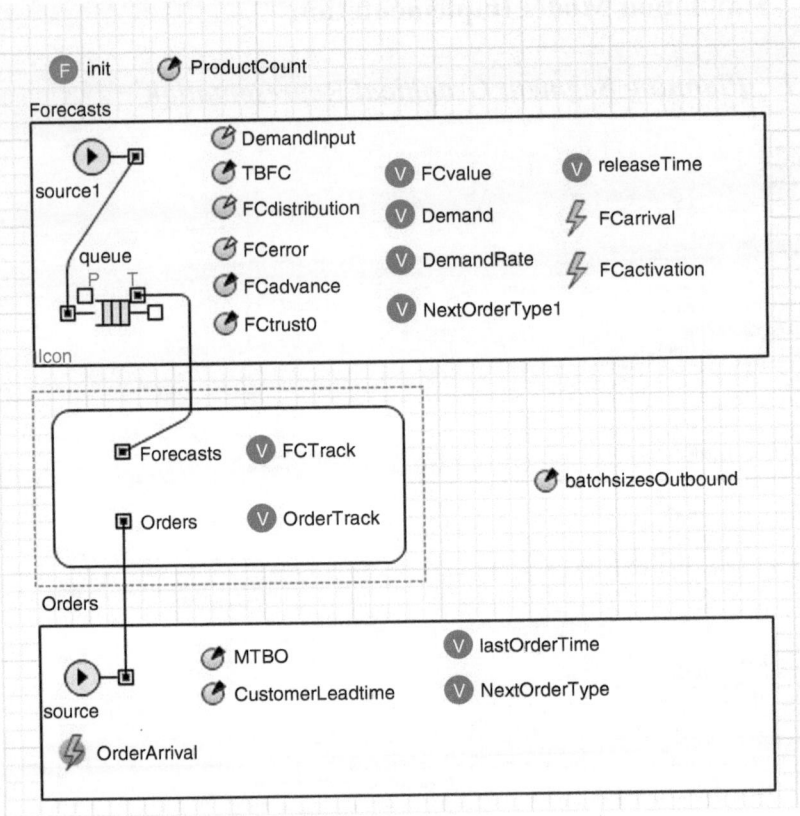

Fig. 8.2 AnyLogic graphical representation of demand model

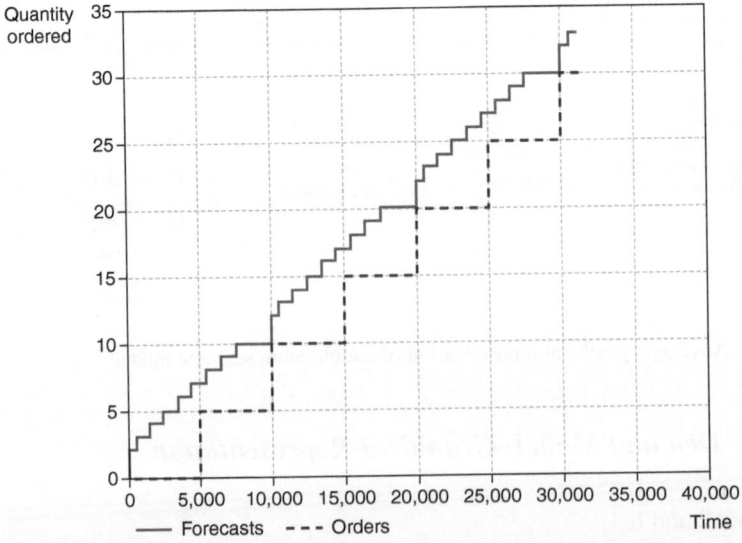

Fig. 8.3 Sample demand model output of cumulated POs (based on forecasts and customer orders)

8.9.3 Java Code Executed at Forecast Arrival (Event 'FCarrival')

```java
//Initialization
int FCBatch=0;
double[] FCorderTimeWindow=new double[100];
boolean done=false;


//Calculation of Demand rate, number of F/C based production orders
//and the time window to release F/C based production orders for every
//product type
for (int z=0;z<ProductCount;z++) {
    FCBatch=
    Math.max(0,(int) round(DemandInput()[z]/batchsizesOutbound[z]));
    Demand[z]=FCBatch*batchsizesOutbound[z];
    FCvalue[z]=(int)max(0,(Demand[z]+Demand[z]*FCerror()[z]));
    FCvalue[z]=(int)round(FCvalue[z]*FCtrust[z]);
    FCvalue[z]=
    (round(FCvalue[z]/batchsizesOutbound[z]))*batchsizesOutbound[z];
    FCorderTimeWindow[z]=
    TBFC/(round(FCvalue[z]/batchsizesOutbound[z])); }

//Trigger the activation of the new demand rate
FCactivation.restart(FCadvance);

//Level the sequence of F/C based production orders, assign release
//times and inject them in a queue where they wait to be released
while (!done){
    done=true;
    for (int z=0;z<ProductCount;z++) {
    NextOrderType1=z;
    if (FCvalue[z]!=0) {
        releaseTime=Math.max(time()+
        (FCvalue[z]/batchsizesOutbound[z])*FCorderTimeWindow[z]-
        FCadvance,0);
        source1.inject(1);
        FCvalue[z]=FCvalue[z]-batchsizesOutbound[z];
        done=false;}
    }
}
```

8.10 PCS Engineering Process: Numerical Optimization of Model Parameters

8.10.1 Setups for Experiments and Illustrations

Tables 8.3, 8.4 and 8.5

Table 8.3 System configuration for the K_i optimization illustration[a]

Category	Parameter	Values K_i optimization
Demand model	Number of products (m)	1
	FC error [m]	{0}
	Demand distribution [m]	{1,000}//Saturated system
	TBFC	10,000
	TBO [m]	{5,000}
	CLT [m]	{0}
	Backlogging	true
	Order batchsize[m]	{20}
Planning segments	Process steps (n)	3
	Transport batchsizes [n]	{20,20,20}
	SCT[n,m]	{(1),(3),(1.5)}
	Transportation [n]	{uniform(20,60), uniform(20,60), uniform(20,60)}
	C/O time [n]	{0,0,0}
	Setup family[n,m]	{(0), (0), (0)}
	Maintenance TTR [n]	{0,0,0}
	Maintenance TBF [n]	{*Inf, Inf, Inf*}
	Scrap PBF [n]	{*Inf, Inf, Inf*}
	Rework PBF [n]	{*Inf, Inf, Inf*}
	Rework TTR [n]	{0,0,0}
	Minor Stops PBF [n]	{exponential(0.01), exponential(0.07), exponential(0.01)}
	Minor Stops TTR [n]	{uniform(6,15), uniform(6,15), uniform(6,15)}
	Breakdown TBF [n]	{uniform(0,500), uniform(0,500), uniform(0,500)}
	Breakdown TTR [n]	{150,150,150}
	Shift Uptime [n]	{960, 1,440, 960}
	Shift Downtime [n]	{480,0,480}
Simulation	Replications	1
	Seed(s)	3
	Model runtime	600,000
	Total work content[m]	{100,100,100}//assumed for productivity calculation

[a]In the following Java[TM] syntax will be used to specify arrays, Inf used for Double/Integer. POSITIVE_INFINITY

Table 8.4 System configuration for basestock allocation experiment

Category	Parameter	Values
Demand model	Number of products (m)	1
	FC error [m]	{0}
	Demand distribution [m]	{uniform(0,600)}
	TBFC	10,000
	TBO [m]	{5,000}
	CLT [m]	{0}
	Backlogging	true
	Order batchsize[m]	{20}
Planning segments	Process steps (n)	3
	Transport batchsizes [n]	{20,20,20}
	SCT[n,m]	{(1),(3),(1.5)}
	Transportation [n]	{uniform(20,60), uniform(20,60), uniform(20,60)}
	C/O time [n]	{0,0,0}
	Setup family[n,m]	{(0), (0), (0)}
	Maintenance TTR [n]	{0,0,0}
	Maintenance TBF [n]	{*Inf, Inf, Inf*}
	Scrap PBF [n]	{*Inf, Inf, Inf*}
	Rework PBF [n]	{*Inf, Inf, Inf*}
	Rework TTR [n]	{0,0,0}
	Minor Stops PBF [n]	{exponential(0.01), exponential(0.07), exponential(0.01)}
	Minor Stops TTR [n]	{uniform(6,15), uniform(6,15), uniform(6,15)}
	Breakdown TBF [n]	{uniform(0,500), uniform(0,500), uniform(0,500)}
	Breakdown TTR [n]	{150,150,150}
	Shift Uptime [n]	{960, 1,440, 960}
	Shift Downtime [n]	{480,0,480}
Simulation	Replications	1
	Seed(s)	3
	Model runtime	600,000
	Utility function parameters	WIP_min=50;WIP_max=350;WIP_weight=0.3; del_lowerBound=0.7;del_upperBound=0.95; del_weight=0.7;
Results	Kanbans*[m]	{4,5}
	FCadvance*	1,600

8.10.2   Influence of Over-Capacity on the MTF Versus MTS Decision

Liberopoulos and Koukoumialos (2005) showed for the EKCS that a bound for K_i exists, exceeding which will not change the optimal choice of S_{ij}. However, for the generic PCS model we propose, over capacity has an influence on the trade-off between MTS and MTF. This is rooted in the nature of MRP to determine order points based on a due date and lead time assumptions. Figure 8.4 shows the basestock development over time in an experiment comparing MTS and MTF in both, a setting

Table 8.5 Case used for FCT_j and S_{ij} optimization illustration

Category	Parameter	Values
Demand model	Number of products (m)	1
	FC error [m]	{−0.2+bernoulli(0.5)*0.4}
	Demand distribution [m]	{bernoulli(0.5)*600}
	TBFC	10,000
	TBO [m]	{5,000}
	CLT [m]	{0}
	Backlogging	true
	Order batchsize[m]	{20}
Planning segments	Process steps (n)	3
	Transport batchsizes [n]	{20,20,20}
	SCT[n,m]	{(1),(3),(1.5)}
	Transportation [n]	{uniform(20,60), uniform(20,60), uniform(20,60)}
	C/O time [n]	{0,0,0}
	Setup family[n,m]	{(0), (0), (0)}
	Maintenance TTR [n]	{0,0,0}
	Maintenance TBF [n]	{Inf, Inf, Inf}
	Scrap PBF [n]	{Inf, Inf, Inf}
	Rework PBF [n]	{Inf, Inf, Inf}
	Rework TTR [n]	{0,0,0}
	Minor Stops PBF [n]	{exponential(0.01), exponential(0.07), exponential(0.01)}
	Minor Stops TTR [n]	{uniform(6,15), uniform(6,15), uniform(6,15)}
	Breakdown TBF [n]	{uniform(0,500), uniform(0,500), uniform(0,500)}
	Breakdown TTR [n]	{150,150,150}
	Shift Uptime [n]	{960, 1,440, 960}
	Shift Downtime [n]	{480,0,480}
Simulation	Replications	1
	Seed(s)	4
	Model runtime	600,000
	Utility function parameters	WIP_min=50;WIP_max=350;WIP_weight=0.3; del_lowerBound=0.7;del_upperBound=0.95; del_weight=0.7;
Results	Kanbans*[m]	{4,5}
	FCadvance*	1,600
	FCT_j*	{0.7}
	S_{ij}*	{7}

with optimal production capacity, and a setting with over capacity. The analysis has been performed in a setting with no demand variability or forecast error.

In a setting with over capacity, the basestock is replenished much faster by the MTS than the MTF policy, and much faster than needed. Thus the average WIP level of the MTS system is inherently higher. In an environment with optimal system capacity (achieved by selecting K_i optimal) and no demand variability and forecast error, MTS and MTF perform approximately equal.

The following experimental setup was used (Table 8.6).

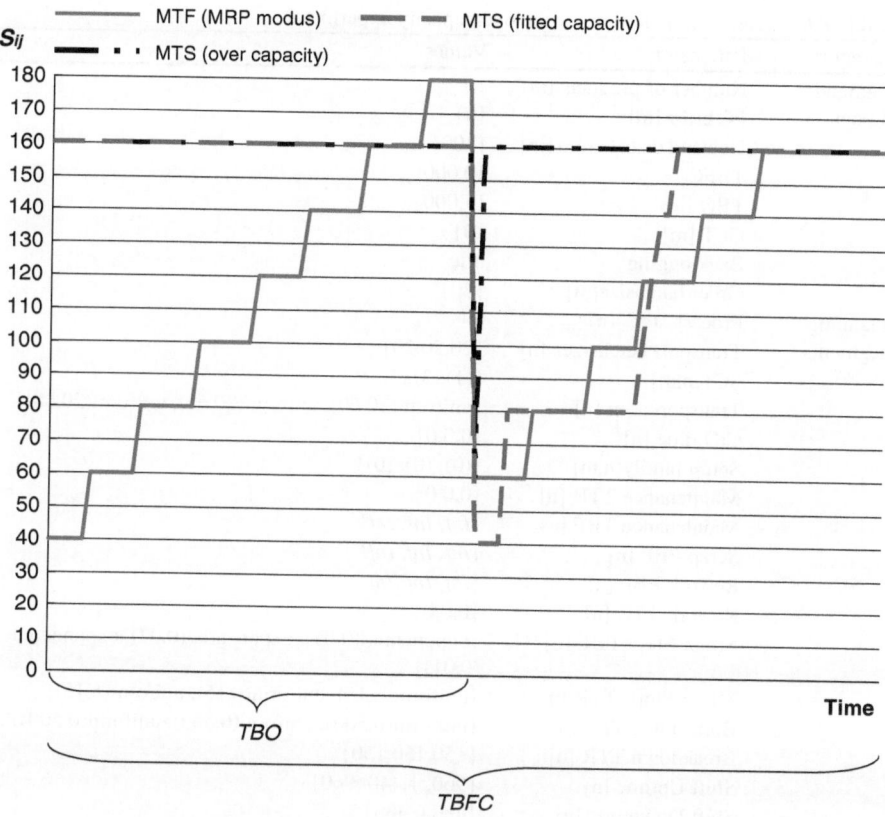

Fig. 8.4 Effect of over-capacity on the MTF versus MTS decision

8.10.3   *Converting Utility Improvements into WIP Reductions*

$$u = w_1\left(1 - \frac{WIP - WIP_{\min}}{WIP_{\max} - WIP_{\min}}\right) + w_2 v(DelPerf) \overset{solve\,for\,WIP}{\Longrightarrow}$$

$$u - w_2 v(DelPerf) = w_1\left(1 - \frac{WIP - WIP_{\min}}{WIP_{\max} - WIP_{\min}}\right) \Rightarrow$$

$$\frac{-u + w_2 v(DelPerf)}{w_1} + 1 = \frac{WIP - WIP_{\min}}{WIP_{\max} - WIP_{\min}} \Rightarrow$$

$$(WIP_{\max} - WIP_{\min}) \cdot \left(\frac{-u + w_2 v(DelPerf)}{w_1} + 1\right) - WIP_{\min} = WIP \Rightarrow$$

$$-\frac{WIP_{\max} - WIP_{\min}}{w_1} u + (WIP_{\max} - WIP_{\min}) \cdot \left(\frac{w_2 v(DelPerf)}{w_1} + 1\right) - WIP_{\min}$$

$$\underset{1st\,deviation}{= WIP} \overset{\frown}{\Rightarrow}$$

$$WIP'(u)du = -\frac{WIP_{\max} - WIP_{\min}}{w_1}$$

Multiplying this factor with a utility increase delivers the corresponding WIP decrease.

Table 8.6 System configuration for the overcapacity experiment

Category	Parameter	Values
Demand model	Number of products (m)	1
	FC error [m]	{0}
	Demand distribution [m]	{300}
	TBFC	10,000
	TBO [m]	{5,000}
	CLT [m]	{0}
	Backlogging	true
	Order batchsize[m]	{20}
Planning segments	Process steps (n)	3
	Transport batchsizes [n]	{20,20,20}
	SCT[n,m]	{(1),(3),(1.5)}
	Transportation [n]	{uniform(20,60), uniform(20,60), uniform(20,60)}
	C/O time [n]	{0,0,0}
	Setup family[n,m]	{(0), (0), (0)}
	Maintenance TTR [n]	{0,0,0}
	Maintenance TBF [n]	{*Inf, Inf, Inf*}
	Scrap PBF [n]	{*Inf, Inf, Inf*}
	Rework PBF [n]	{*Inf, Inf, Inf*}
	Rework TTR [n]	{0,0,0}
	Minor Stops PBF [n]	{exponential(0.01), exponential(0.07), exponential (0.01)}
	Minor Stops TTR [n]	{uniform(6,15), uniform(6,15), uniform(6,15)}
	Breakdown TBF [n]	{uniform(0,500), uniform(0,500), uniform(0,500)}
	Breakdown TTR [n]	{150,150,150}
	Shift Uptime [n]	{960, 1,440, 960}
	Shift Downtime [n]	{480,0,480}
Simulation	Replications	1
	Seed(s)	1
	Model runtime	600,000
	Utility function parameters	WIP_min=50;WIP_max=350;WIP_weight=0.3; del_lowerBound=0.7;del_upperBound=0.95; del_weight=0.7;
Results	Kanbans*[m] – over capacity	{6,5}
	Kanbans*[m] – opt capacity	{2,2}
	FCadvance*	1,600

8.11   Investigation of the Push/Pull Integration

8.11.1   *Graphical Representation of Experimental Setup in Anylogic*

Figure 8.5, Tables 8.7 and 8.8

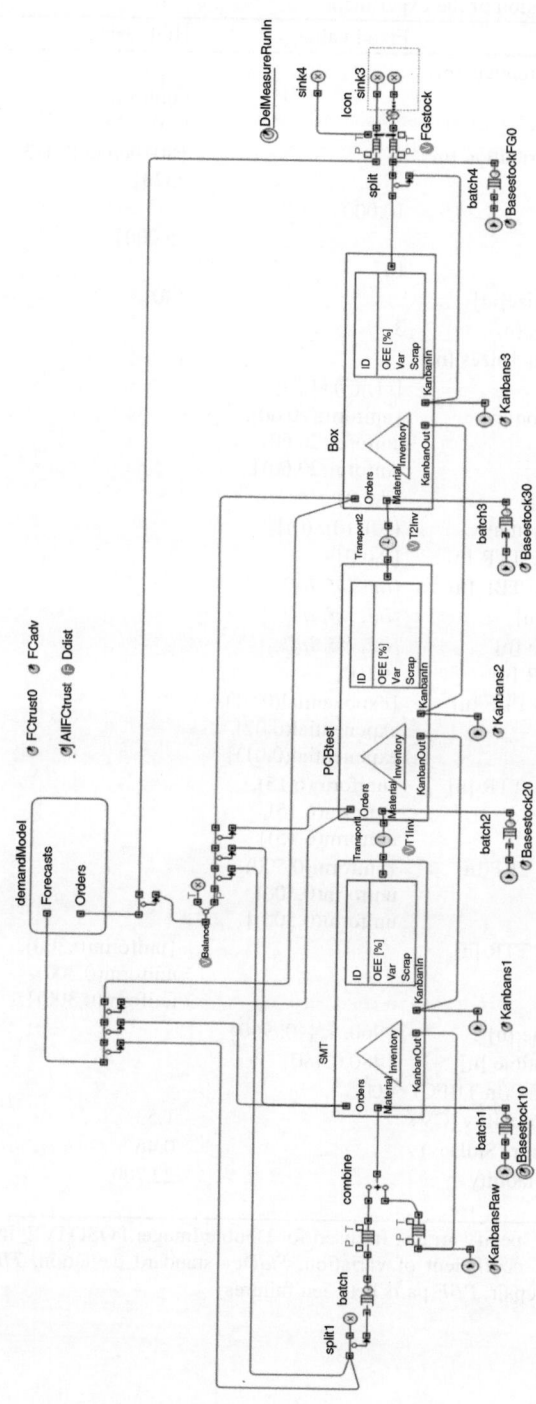

Fig. 8.5 AnyLogic production system model

Table 8.7 Parameterization of the experiment

Category	Parameter	Fixed value	High value	Low value
Demand model	Number of products (m)	1		
	FC error [m]		{uniform (−0.8,0.8)	{0}
	Demand distribution [m]		{80+bernoulli(0.3) *320}	{400}
	TBFC	10,000		
	TBO [m]		{5,000}	{1,000}
	CLT [m]	{0}		
	Order batchsize[m]		{40}	{10}
Planning segments	Process steps (n)	3		
	Transport batchsizes [n]		{40,40,40}	{10,10,10}
	SCT[n,m]	{(1),(3),(1.5)}		
	Transportation [n]	{uniform(20,60), uniform(20,60), uniform(20,60)}		
	C/O time [n]	{0,0,0}		
	Setup family[n,m]	{(0), (0), (0)}		
	Maintenance TTR [n]	{0,0,0}		
	Maintenance TBF [n]	*{Inf, Inf, Inf}*		
	Scrap PBF [n]	*{Inf, Inf, Inf}*		
	Rework PBF [n]	*{Inf, Inf, Inf}*		
	Rework TTR [n]	{0,0,0}		
	Minor Stops PBF [n]	{exponential(0.01), exponential(0.07), exponential(0.01)}		
	Minor Stops TTR [n]	{uniform(6,15), uniform(6,15), uniform(6,15)}		
	Breakdown TBF [n]	{uniform(0,500), uniform(0,500), uniform(0,500)}		
	Breakdown TTR [n]		{uniform(0,300), uniform(0,300), uniform(0,300)}	{150,150,150}
	Shift Uptime [n]	{960, 1,440, 960}		
	Shift Downtime [n]	{480,0,480}		
Derived factors	Peak demand (in TBFC)	400		
	Demand variability (CV)		1.53	0
	Forecast error (StdDev)		0.46	0
	Process variability (VMPS)		22,700	16,700

Java^{TM} syntax used to specify arrays; Inf used for Double/Integer.POSITIVE_INFINITY

C/O change over, *CV* coefficient of variation, *StdDev* standard deviation, *TBF* time between failures, *TTR* time to repair, *PBF* parts between failures

Table 8.8 Experiment result table

Standard order	Run Order	Center Point	Blocks	TBO	Batch size	Demvar. (CV)	VMPS	F/C error (CV)	Seed	FCadvance	FCT*	S*	Sreach	PureS	S %*	Utility*	Utility-Irrpr-MTF	Utility-Impr-MTS
1	1	1	1	1,000	10	0.00	16,700	0.46	1	1,200	0.0	8	2,000	8	1.00	0.96	1.63	0.00
2	2	1	1	5,000	10	0.00	16,700	0.00	1	1,300	1.0	0	0	20	0.00	0.89	0.00	0.06
3	3	1	1	1,000	40	0.00	16,700	0.00	69	1,500	1.0	0	0	3	0.00	0.97	0.00	0.03
4	4	1	1	5,000	40	0.00	16,700	0.46	1	2,000	0.0	5	5,000	5	1.00	0.83	*	0.00
5	5	1	1	1,000	10	1.53	16,700	0.00	1	1,300	1.0	0	0	8	0.00	1.05	0.00	0.09
6	6	1	1	5,000	10	1.53	16,700	0.46	1	2,100	0.6	4	1,000	20	0.20	0.93	0.13	0.14
7	7	1	1	1,000	40	1.53	16,700	0.46	1	1,500	0.6	2	2,000	3	0.67	0.97	1.19	0.04
8	8	1	1	5,000	40	1.53	16,700	0.46	1	1,500	0.6	2	2,000	3	0.67	0.97	1.19	0.04
9	9	1	1	1,000	10	0.00	22,700	0.00	1	1,200	1.0	0	0	0	0.00	1.03	0.00	0.26
10	10	1	1	5,000	10	0.00	22,700	0.46	1	1,500	0.0	20	5,000	20	1.00	0.84	2.94	0.00
11	11	1	1	1,000	40	0.00	22,700	0.46	1	1,700	0.0	3	30,000	3	1.00	0.94	1.23	0.00
12	12	1	1	5,000	40	0.00	22,700	0.00	69	1,900	1.0	0	0	5	0.00	0.87	0.00	0.05
13	13	1	1	1,000	10	1.53	22,700	0.46	1	1,400	0.5	5	1,250	8	0.63	0.99	0.39	0.02
14	14	1	1	5,000	10	1.53	22,700	0.00	1	2,000	1.0	0	0	20	0.00	1.02	0.00	0.24
15	15	1	1	1,000	40	1.53	22,700	0.00	69	1,700	1.0	0	0	2	0.00	1.04	0.00	0.09
16	16	1	1	5,000	40	1.53	22,700	0.46	1	2,700	0.7	1	1,000	5	0.20	0.94	0.28	0.15
17	17	1	1	1,000	10	0.00	16,700	0.46	5	1,200	0.0	8	2,000	8	1.00	0.96	0.56	0.00
18	18	1	1	5,000	10	0.00	16,700	0.00	5	1,300	1.0	0	0	20	0.00	0.84	0.00	0.06
19	19	1	1	1,000	40	0.00	16,700	0.00	5	1,500	1.0	0	0	3	0.00	0.97	0.00	0.02
20	20	1	1	5,000	40	0.00	16,700	0.46	5	2,000	0.0	5	50,000	5	1.00	0.83	0.99	0.00
21	21	1	1	1,000	10	1.53	16,700	0.00	5	1,300	1.0	0	0	8	0.00	1.06	0.00	0.08
22	22	1	1	5,000	10	1.53	16,700	0.46	5	2,100	0.6	4	1,000	20	0.20	0.91	0.32	0.10
23	23	1	1	1,000	40	1.53	16,700	0.46	5	1,500	0.4	2	2,000	2	1.00	0.97	0.24	0.10
24	24	1	1	5,000	40	1.53	16,700	0.00	5	2,900	1.0	0	0	5	0.00	1.02	0.00	0.24
25	25	1	1	1,000	10	0.00	22,700	0.46	5	1,200	1.0	0	0	0	0.00	0.96	0.00	0.00
26	26	1	1	5,000	10	0.00	22,700	0.46	5	1,500	0.0	20	5,000	20	1.00	0.84	*	0.00
27	27	1	1	1,000	40	0.00	22,700	0.46	5	1,700	0.0	3	3,000	3	1.00	0.94	2.34	0.00
28	28	1	1	5,000	40	0.00	22,700	0.00	5	1,900	1.0	0	0	5	0.00	0.87	0.00	0.05
29	29	1	1	1,000	10	1.53	22,700	0.46	5	1,200	0.5	5	1,250	8	0.63	0.98	0.17	0.00
30	30	1	1	5,000	10	1.53	22,700	0.00	5	2,000	1.0	0	0	20	0.00	1.02	0.00	0.25

(continued)

Table 8.8 (continued)

Standard order	Run Order	Center Point	Blocks	TBO	Batch size	Demvar. (CV)	VMPS	F/C error (CV)	Seed	FCadvance	FCT*	S*	Sreach	PureS	S %*	Utility*	Utility-Irrpr-MTF	Utility-Impr-MTS
31	31	1	1	1,000	40	1.53	22,700	0.00	5	1,700	1.0	0	0	2	0.00	1.04	0.00	0.10
32	32	1	1	5,000	40	1.53	22,700	0.46	5	2,700	0.6	1	1,000	5	0.20	0.99	0.28	0.21
33	33	1	1	1,000	10	0.00	16,700	0.46	69	1,200	0.0	10	2,500	10	1.00	0.96	1.07	0.00
34	34	1	1	5,000	10	0.00	16,700	0.00	69	1,300	1.0	0	0	20	0.00	0.89	0.00	0.06
35	35	1	1	1,000	40	0.00	16,700	0.00	1	1,500	1.0	0	0	3	0.00	0.96	0.00	0.02
36	36	1	1	5,000	40	0.00	16,700	0.46	69	2,000	0.0	5	5,000	5	1.00	0.83	4.30	0.00
37	37	1	1	1,000	10	1.53	16,700	0.00	69	1,300	1.0	0	0	8	0.00	1.05	0.00	0.07
38	38	1	1	5,000	10	1.53	16,700	0.46	69	2,100	0.5	4	1,000	20	0.20	0.92	0.05	0.12
39	39	1	1	1,000	40	1.53	16,700	0.46	69	1,500	0.6	2	2,000	2	1.00	0.98	0.85	0.02
40	40	1	1	5,000	40	1.53	16,700	0.00	69	2,900	1.0	0	0	5	0.00	1.04	0.00	0.27
41	41	1	1	1,000	10	0.00	22,700	0.00	69	1,200	1.0	0	0	0	0.00	0.96	0.00	0.00
42	42	1	1	5,000	10	0.00	22,700	0.46	69	1,500	0.0	20	5,000	20	1.00	0.83	2.64	0.00
43	43	1	1	1,000	40	0.00	22,700	0.46	69	1,700	0.0	3	3,000	3	1.00	0.94	0.39	0.00
44	44	1	1	5,000	40	0.00	22,700	0.00	1	1,900	1.0	0	0	5	0.00	0.87	0.00	0.05
45	45	1	1	1,000	10	1.53	22,700	0.46	69	1,400	0.6	5	1,250	8	0.63	0.98	0.04	0.00
46	46	1	1	5,000	10	1.53	22,700	0.00	69	2,000	1.0	0	0	20	0.00	1.03	0.00	0.25
47	47	1	1	1,000	40	1.53	22,700	0.00	1	1,700	1.0	0	0	3	0.00	1.04	0.00	0.11
48	48	1	1	5,000	40	1.53	22,700	0.46	69	2,700	0.5	1	1,000	5	0.20	0.96	0.28	0.17

8.11.2   Experiments and Results to Determine Influencing Factors for Upstream Control

General Linear model for FCT*

```
Estimated Effects and Coefficients for FCT* (coded units)

Term                        Effect     Coef    SE Coef        T       P
Constant                             0.6396   0.006250   102.33   0.000
TBO                         0.0125   0.0063   0.006250     1.00   0.325
Batchsize                   0.0042   0.0021   0.006250     0.33   0.741
Demand_var                  0.2792   0.1396   0.006250    22.33   0.000
Process_var                 0.0042   0.0021   0.006250     0.33   0.741
FC_err                     -0.7208  -0.3604   0.006250   -57.67   0.000
TBO*Batchsize               0.0042   0.0021   0.006250     0.33   0.741
TBO*Demand_var              0.0125   0.0062   0.006250     1.00   0.325
TBO*Process_var             0.0042   0.0021   0.006250     0.33   0.741
TBO*FC_err                  0.0125   0.0062   0.006250     1.00   0.325
Batchsize*Demand_var        0.0042   0.0021   0.006250     0.33   0.741
Batchsize*Process_var       0.0125   0.0062   0.006250     1.00   0.325
Batchsize*FC_err            0.0042   0.0021   0.006250     0.33   0.741
Demand_var*Process_var      0.0042   0.0021   0.006250     0.33   0.741
Demand_var*FC_err           0.2792   0.1396   0.006250    22.33   0.000
Process_var*FC_err          0.0042   0.0021   0.006250     0.33   0.741


S = 0.0433013   R-Sq = 99.27%   R-Sq(adj) = 98.92%


Analysis of Variance for FCT* (coded units)

Source              DF   Seq SS   Adj SS   Adj MS        F       P
Main Effects         5  7.17271  7.17271  1.43454   765.09   0.000
2-Way Interactions  10  0.94208  0.94208  0.09421    50.24   0.000
Residual Error      32  0.06000  0.06000  0.00187
  Pure Error        32  0.06000  0.06000  0.00188
Total               47  8.17479


Unusual Observations for FCT*

Obs  StdOrder  FCT*      Fit   SE Fit  Residual  St Resid
 16        16  0.70000  0.60000  0.02500  0.10000    2.83R
 23        23  0.40000  0.53333  0.02500 -0.13333   -3.77R
 48        48  0.50000  0.60000  0.02500 -0.10000   -2.83R

R denotes an observation with a large standardized residual.


Estimated Coefficients for FCT* using data in uncoded units

Term                             Coef
Constant                      1.12066
TBO                        -1.17014E-05
Batchsize                  -0.00308333
Demand_var                  -0.0346768
Process_var                -5.20833E-06
FC_err                        -2.28925
TBO*Batchsize               6.94444E-08
TBO*Demand_var              4.08497E-06
TBO*Process_var             3.47222E-10
TBO*FC_err                  1.35870E-05
Batchsize*Demand_var        0.000181554
Batchsize*Process_var       1.38889E-07
Batchsize*FC_err            0.00060386
Demand_var*Process_var      9.07771E-07
Demand_var*FC_err             0.793312
Process_var*FC_err          3.01932E-06
```

All following residuals plots for *FCT** show no abnormalities that would require further analysis.

Figure 8.6

Regression Analysis: S%* versus FCT*

```
The regression equation is
S%* = 0.984 - 0.994 FCT*


46 cases used. 2 cases contain missing values


Predictor        Coef   SE Coef        T       P
Constant      0.98393   0.02758    35.68   0.000
FCT*     -0.99380   0.03581   -27.75   0.000


S = 0.101999   R-Sq = 94.6%   R-Sq(adj) = 94.5%


Analysis of Variance

Source           DF        SS        MS        F       P
Regression        1    8.0138    8.0138   770.27   0.000
Residual Error   44    0.4578    0.0104
Total            45    8.4715


Unusual Observations

Obs    FCT*        S%*      Fit  SE Fit   Residual   St Resid
  7     0.60    0.6700   0.3877   0.0151     0.2823       2.80R
 38     0.50    0.2000   0.4870   0.0159    -0.2870      -2.85R
 45     0.60    0.6250   0.3877   0.0151     0.2373       2.35R
 48     0.50    0.2000   0.4870   0.0159    -0.2870      -2.85R

R denotes an observation with a large standardized residual.
```

8.11.3   Experiments and Results to Determine Closed-Form Solutions for **FCT*** and **S%***

Tables 8.9 and 8.10

General Linear Model: FCT* versus D_var; FC_err

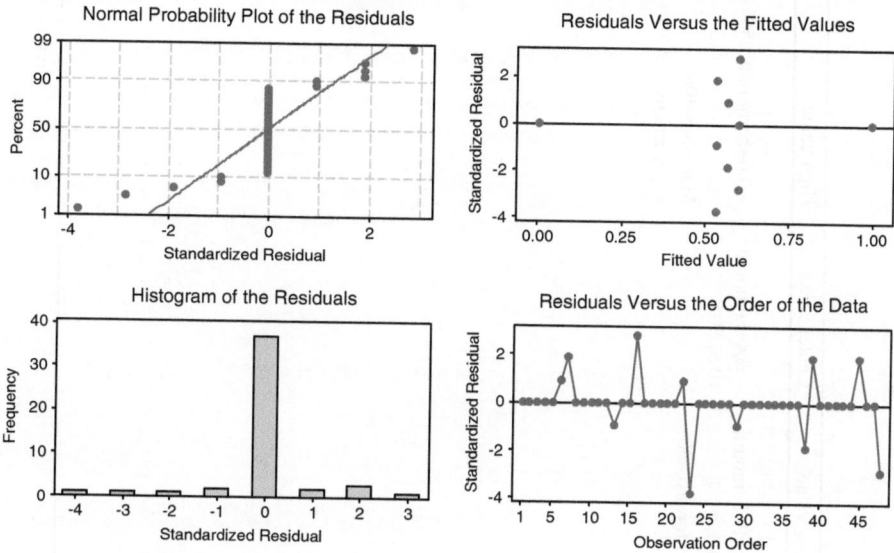

Fig. 8.6 Residual plots for FCT* (Experiment in Sect. 5.1)

```
Factor   Type    Levels  Values
D_var    fixed      2    2; 3
FC_err   fixed      3    1; 2; 3


Analysis of Variance for FCT*. using Adjusted SS for Tests

Source          DF     Seq SS     Adj SS     Adj MS       F      P
D_var            1   0.011468   0.001667   0.001667    1.25  0.290
FC_err           2   0.131698   0.114333   0.057167   42.88  0.000
D_var*FC_err     2   0.001000   0.001000   0.000500    0.37  0.697
Error           10   0.013333   0.013333   0.001333
Total           15   0.157500


S = 0.0365148   R-Sq = 91.53%   R-Sq(adj) = 87.30%


Unusual Observations for FCT*

Obs   FCT*       Fit      SE Fit   Residual  St Resid
  1  0.600000  0.600000  0.036515  0.000000       *  X
 18  0.900000  0.833333  0.021082  0.066667     2.24  R
 22  0.500000  0.566667  0.021082 -0.066667    -2.24  R

R denotes an observation with a large standardized residual.
X denotes an observation whose X value gives it large influence.
```

Figure 8.7

Table 8.9 Parameterization of the experiment

Category	Parameter	Fixed value	Low value	Center value	High value
Demand model	Number of products (m)	1			
	FC error [m]		−0.2+bernoulli (0.5) *0.4	−0.4+bernoulli (0.5) *0.8	−0.6+bernoulli(0.5) *1.2
	Demand distribution [m]		uniform(0,400)		Math.min(400, exponential(1)*100)
	TBFC	10,000			
	TBO [m]	{5,000}			
	CLT [m]	{0}			
	Order batchsize[m]	{20,20,20}			
Planning segments	Process steps (n)	3			
	Transport batchsizes [n]	{20,20,20}			
	SCT[n,m]	{(1),(3),(1.5)}			
	Transportation [n]	{uniform(20,60), uniform(20,60), uniform (20,60)}			
	C/O time [n]	{0,0,0}			
	Setup family[n,m]	{(0), (0), (0)}			
	Maintenance TTR [n]	{0,0,0}			
	Maintenance TBF [n]	{Inf, Inf, Inf}			
	Scrap PBF [n]	{Inf, Inf, Inf}			
	Rework PBF [n]	{Inf, Inf, Inf}			
	Rework TTR [n]	{0,0,0}			
	Minor Stops PBF [n]	{exponential(0.01), exponential(0.07), exponential(0.01)}			
	Minor Stops TTR [n]	{uniform(6,15), uniform(6,15), uniform (6,15)}			

(continued)

Table 8.9 (continued)

Category	Parameter	Fixed value	Low value	Center value	High value
	Breakdown TBF [n]	{uniform(0,500), uniform(0,500), uniform (0,500)}			
	Breakdown TTR [n]	{150,150,150}			
	Shift Uptime [n]	{960, 1,440, 960}			
	Shift Downtime [n]	{480,0,480}			
Derived factors	Peak demand (in TBFC)	400			
	Demand variability (CV)		0.58		1
	Forecast error (Exp. deviation in %, '*a*')		0.2	0.4	0.6
	Process variability (VMPS)	16,700			

Table 8.10 Experiment result table

Standard order	Run Order	Pt Type	Blocks	Dem.var (CV)	F/C error (a)	Seed	FCadvance	FCT*	S*	Sreach	PureS	S %*	Utility*	Utility-Impr-MTF	Utility Impr-MTS
1	1	1	1	0.58	0.6	1	1500	0.6	5	5000	8	0.63	0.95	0.76	0.01
2	2	1	1	1.00	0.6	8	2200	0.6	4	8000	7	0.57	0.99	0.42	0.04
3	3	1	1	1.00	0.6	9	2200	0.6	4	8000	7	0.57	0.99	0.27	0.04
4	4	1	1	1.00	0.2	2	2200	0.8	2	4000	6	0.33	1.01	0.04	0.05
5	5	1	1	1.00	0.2	3	2200	0.8	2	4000	7	0.28	1.02	0.26	0.07
6	6	1	1	0.58	0.6	2	1500	*	*	*	*	*	*	*	*
7	7	1	1	1.00	0.2	55	2200	0.8	2	4000	8	0.25	1.02	0.12	0.10
8	8	1	1	1.00	0.4	1	2200	0.7	3	6000	6	0.50	1.00	0.73	0.04
9	9	1	1	0.58	0.2	1	1500	0.8	2	2000	8	0.25	0.98	0.12	0.04
10	10	1	1	0.58	0.4	1	1500	0.7	4	4000	8	0.50	0.96	0.11	0.03
11	11	1	1	0.100	0.4	5	2200	0.7	3	6000	7	0.42	1.00	0.74	0.08
12	12	1	1	0.58	0.2	2	1500	0.9	2	2000	8	0.25	0.98	0.07	0.05
13	13	1	1	0.58	0.6	2	1500	*	*	*	*	*	*	*	*
14	14	1	1	0.58	0.2	69	1600	0.8	3	3000	8	0.38	0.97	0.18	0.04
15	15	1	1	1.00	0.6	10	2200	0.5	3	6000	8	0.38	1.01	0.79	0.09
16	16	1	1	0.58	0.4	2	1500	0.7	4	4000	8	0.50	0.97	0.35	0.04
17	17	1	1	1.00	0.4	69	2200	0.7	3	6000	7	0.42	1.00	1.11	0.06
18	18	1	1	0.58	0.4	3	1500	0.7	4	4000	8	0.50	0.96	(1.96)	0.03

A "*" in the table above indicates that the optimization did not yield an unambiguous optimum and the replication has therefore been ignored

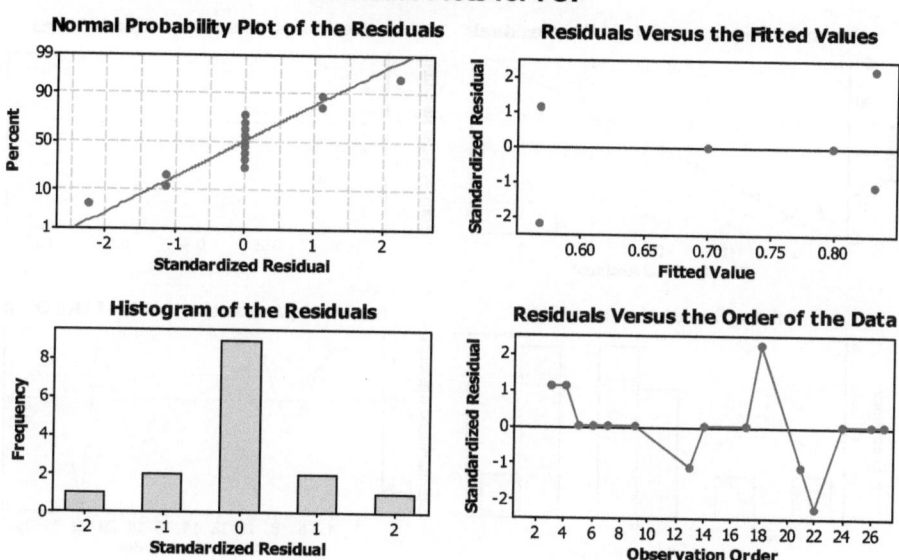

Fig. 8.7   Residual plots for FCT* linear model (Experiment in Sect. 5.2)

General Linear Model: S%* versus D_var; FC_err

```
Factor    Type    Levels   Values
D_var     fixed      2     2; 3
FC_err    fixed      3     1; 2; 3
```

Analysis of Variance for S%*. using Adjusted SS for Tests

```
Source        DF     Seq SS     Adj SS     Adj MS        F       P
D_var          1   0.007840   0.014832   0.014832     3.33   0.101
FC_err         2   0.149072   0.148539   0.074270    16.69   0.001
D_var*FC_err   2   0.004278   0.004278   0.002139     0.48   0.633
Error          9   0.040050   0.040050   0.004450
Total         14   0.201240
```

S = 0.0667083   R-Sq = 80.10%   R-Sq(adj) = 69.04%

Unusual Observations for S%*

```
Obs       S%*        Fit     SE Fit    Residual   St Resid
  1   0.630000   0.630000   0.066708   0.000000       * X
 22   0.380000   0.506667   0.038514  -0.126667    -2.33 R
```

R denotes an observation with a large standardized residual.

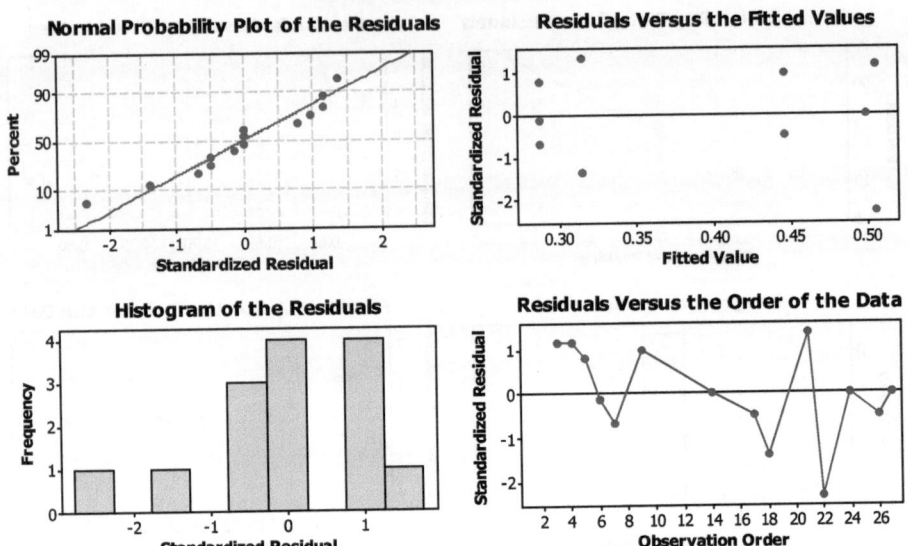

Fig. 8.8 Residual plots for S%* linear model

Figure 8.8

General Linear Model: S* versus D_var; FC_err

```
Factor   Type   Levels  Values
D_var    fixed      2   2; 3
FC_err   fixed      3   1; 2; 3


Analysis of Variance for S*. using Adjusted SS for Tests

Source        DF   Seq SS    Adj SS   Adj MS      F      P
D_var          1   1.1468    2.6667   2.6667   20.00   0.001
FC_err         2  10.6698   10.7333   5.3667   40.25   0.000
D_var*FC_err   2   0.6000    0.6000   0.3000    2.25   0.156
Error         10   1.3333    1.3333   0.1333
Total         15  13.7500


S = 0.365148   R-Sq = 90.30%   R-Sq(adj) = 85.45%


Unusual Observations for S*

Obs      S*        Fit    SE Fit  Residual  St Resid
  1  5.00000  5.00000   0.36515   0.00000       *  X
 21  3.00000  2.33333   0.21082   0.66667     2.24  R
 22  3.00000  3.66667   0.21082  -0.66667    -2.24  R

R denotes an observation with a large standardized residual.
X denotes an observation whose X value gives it large influence.
```

Regression Analysis: FCT* versus Exp_FCT

```
The regression equation is
FCT* = - 0.145 + 1.17 Exp_FCT


16 cases used. 11 cases contain missing values


Predictor          Coef   SE Coef        T       P
Constant       -0.14500   0.08453    -1.72   0.108
Exp_FCT          1.1667    0.1143    10.20   0.000


S = 0.0365148    R-Sq = 88.1%    R-Sq(adj) = 87.3%


Analysis of Variance

Source            DF        SS        MS       F       P
Regression         1   0.13883   0.13883  104.13   0.000
Residual Error    14   0.01867   0.00133
Total             15   0.15750


Unusual Observations

Obs   Exp_FCT   FCT*       Fit    SE Fit   Residual   St Resid
 18     0.830   0.90000  0.82333  0.01419   0.07667      2.28R
 22     0.630   0.50000  0.59000  0.01508  -0.09000     -2.71R

R denotes an observation with a large standardized residual.
```

Regression Analysis: S* numerical versus S* calculated with upper bound

```
The regression equation is
S* = 0.952 + 0.656 Exp_BS_UB


16 cases used. 11 cases contain missing values


Predictor     Coef   SE Coef     T      P
Constant    0.9520    0.3727   2.55   0.023
Exp_BS_UB   0.6560    0.1057   6.21   0.000


S = 0.511580    R-Sq = 73.4%    R-Sq(adj) = 71.4%


Analysis of Variance

Source           DF      SS      MS       F      P
Regression        1   10.086  10.086   38.54  0.000
Residual Error   14    3.664   0.262
Total            15   13.750


Unusual Observations

Obs   Exp_BS_UB     S*    Fit  SE Fit  Residual  St Resid
 22        5.00  3.000  4.232   0.219    -1.232     -2.67R

R denotes an observation with a large standardized residual.
```

Regression Analysis: S* numerical versus S* arithmetic rounding

```
The regression equation is
S* = 1.45 + 0.596 Exp_BS_rounded


16 cases used. 11 cases contain missing values


Predictor            Coef   SE Coef      T       P
Constant           1.4485    0.4138   3.50   0.004
Exp_BS_rounded     0.5961    0.1356   4.40   0.001


S = 0.642382    R-Sq = 58.0%    R-Sq(adj) = 55.0%


Analysis of Variance

Source              DF       SS       MS       F       P
Regression           1    7.9728   7.9728   19.32   0.001
Residual Error      14    5.7772   0.4127
Total               15   13.7500


Unusual Observations

Obs   Exp_BS_rounded      S*     Fit   SE Fit   Residual   St Resid
 22             5.00   3.000   4.429    0.337     -1.429      -2.61R

R denotes an observation with a large standardized residual.
```

8.11.4   Extension to Arbitrary Forecast Error Distributions

For the following considerations, the assumptions $E(FCerr)=0$ (bias removed) and $q_{0.5}(FCerr) = 0$ will be upheld first, and the relaxation of the second assumption be discussed afterwards. The closed-form parameterization of the hybrid control approach has been constructed and empirically validated based on one, artificially constructed, probability distribution. In the following, the validity of the proposed closed-form parameterization will be tested with two more theoretical probability distributions that fulfill $E(FCerr)=0$ and $q_{0.5}(FCerr) = 0$. Due to their easy handling and suitability for practical applications, the uniform– and the triangular distribution are chosen.

To be able to test the presented approach against the numerical results, the expression $E(|FCerr|)$ needs to be computed for each distribution. For the

Fig. 8.9 Probability density functions of the *triangular* distribution

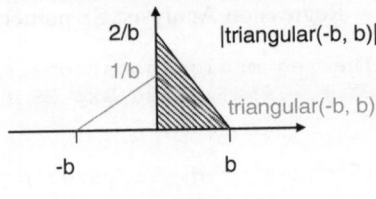

Fig. 8.10 Probability density functions of the *uniform* distribution

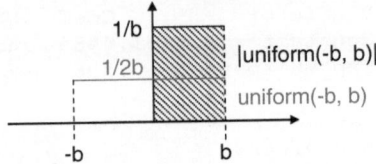

hypothetical, Bernoulli based distribution with parameter a, $E(|FCerr|)=a$ has already been shown in (5.2).

For the triangular distribution, Fig. 8.9 displays the probability density functions (PDF) for a *triangular$(-b,b)$* and a *|triangular$(-b,b)$|* distributed random variable.

The PDF of the *|triangular$(-b,b)$|* distribution equals

$$f(x) = \frac{2}{b} - \frac{2}{b^2}x \tag{8.7}$$

By integration, $E(|FCerr|)$ can be computed as

$$E(|FCerr|) = \int_0^b f(x) \cdot x \, dx = \int_0^b \left(\frac{2}{b}x - \frac{2}{b^2}x^2\right) dx = \left[\frac{1}{b}x^2 - \frac{2}{3b^2}x^3\right]_0^b = \frac{b}{3} \tag{8.8}$$

The same approach can be taken for the uniform distribution.
Figure 8.10
The PDF of the *|uniform$(-b,b)$|* distribution equals

$$f(x) = \frac{1}{b} \tag{8.9}$$

$$E(|FCerr|) = \int_0^b f(x) \cdot x \, dx = \int_0^b \frac{1}{b} \cdot x \, dx = \left[\frac{1}{2b}x^2\right]_0^b = \frac{b}{2} \tag{8.10}$$

andfollows. Please note, that in the distributions above, $b \in [0, 1]$ holds due to the definition of the forecast error in Sect. 3.4.1.

The experiments are executed based on the previously used production system setup which is summarized in the following Table 8.11.

Table 8.11 Experimental setup for the extension to arbitrary forecast error distributions

Category	Parameter	Value
Demand model	Number of products (m)	1
	FC error [m]	Varied, see text
	Demand distribution [m]	Math.min(400,exponential(1)*100)
	TBFC	10,000
	TBO [m]	{5,000}
	CLT [m]	{0}
	Order batchsize[m]	{20,20,20}
Planning segments	Process steps (n)	3
	Transport batchsizes [n]	{20,20,20}
	SCT[n,m]	{(1),(3),(1.5)}
	Transportation [n]	{uniform(20,60), uniform(20,60), uniform(20,60)}
	C/O time [n]	{0,0,0}
	Setup family[n,m]	{(0), (0), (0)}
	Maintenance TTR [n]	{0,0,0}
	Maintenance TBF [n]	{Inf, Inf, Inf}
	Scrap PBF [n]	{Inf, Inf, Inf}
	Rework PBF [n]	{Inf, Inf, Inf}
	Rework TTR [n]	{0,0,0}
	Minor Stops PBF [n]	{exponential(0.01), exponential(0.07), exponential(0.01)}
	Minor Stops TTR [n]	{uniform(6,15), uniform(6,15), uniform(6,15)}
	Breakdown TBF [n]	{uniform(0,500), uniform(0,500), uniform(0,500)}
	Breakdown TTR [n]	{150,150,150}
	Shift Uptime [n]	{960, 1,440, 960}
	Shift Downtime [n]	{480,0,480}
Derived factors	Peak demand (in TBFC)	400
	Demand variability (CV)	1
	Forecast error (StdDev)	Varied
	Process variability (VMPS)	16,700

Table 8.12 Comparison of analytical and numerical FCT^* derivation for different distributions

Distribution	Analytical derivation		Numerical derivation			
	$E(|FCerr|)$	FCT^*	FCT^*	Repl. 1	Repl. 2	Repl. 3
triangular(−0.6,0.6)	0.20	0.83	0.83	0.8	0.9	0.8
triangular(−0.9,0.9)	0.30	0.77	0.77	0.8	0.8	0.7
uniform(−0.4,0.4)	0.20	0.83	0.83	0.9	0.8	0.8
uniform(−0.7,0.7)	0.35	0.74	0.70	0.7	0.7	0.7

The results for FCT^* that are derived analytically using the expressions for $E(|FCerr|)$ calculated above are compared with numerically derived results. Two uniform and two triangular distributions are tested. The numerical determination of FCT* is performed over three replications (Repl.). Table 8.12 shows the results of the test.

Table 8.13 Comparison of analytical and numerical FCT* derivation for distributions with a median different from zero

Distribution	Analytical derivation		Numerical derivation			
	FCT* via $E(|FCerr|)$	FCT* via alternative	FCT*	Repl. 1	Repl. 2	Repl. 3
$-0.3 + bernoulli\left(\frac{1}{3}\right) \cdot 0.9$	0.71	0.63	0.60	0.6	0.6	0.6
$-0.6 + bernoulli\left(\frac{2}{3}\right) \cdot 0.9$	0.71	0.63	0.56	0.6	0.6	0.5

The result clearly confirms that the approach can also be applied with both, the triangular and uniform distribution. Due to the very accurate result, a statistical analysis is omitted.

In practical applications, $E(|FCerr|)$ can also be directly estimated from the forecast error data. The derivation of forecast error data has been defined in (3.26). Based on this and on a given data set with T data points, an estimator for $E(|FCerr|)$ is

$$E(|FCerr|) = \frac{1}{T} \sum_{t=1}^{T} \left| fcerr_t^{relNoBias} \right| \tag{8.11}$$

Next, it is examined what happens if the assumption $q_{0.5}(FCerr) = 0$ is relaxed. However, for most practical applications, this case should be of minor importance. For most cases, the median of the observed forecast errors is so close to zero that the effect of a potential deviation on $FCT*$ is negligible. Two forecast error distributions are constructed, again based on the Bernoulli distribution. One with $q_{0.5}(FCerr) > 0$ and one with $q_{0.5}(FCerr) < 0$. $E(FCerr) = 0$ still needs to hold. The choice of $FCT*$ is determined numerically with three replications. A similar production setup as in the previous experiment is used.[1] The results are summarized in the following (Table 8.13).

As we can see, the proposed approach via $E(|FCerr|)$ seems to be invalid in this environment. However, an alternative calculation approach, which again gets closer to the numerically determined values, can be proposed. The alternative approach only considers positive forecast error deviations. Let POS be the set of positive deviations due to forecast error (either observations from actual data or sampled from a probability distribution), which is a subset from T total observations.

$$POS = \left\{ t = 1, ..., T \mid fcerror_t^{relNoBias} \geq 0 \right\} \tag{8.12}$$

---

[1] Only the demand distribution is changed to {bernoulli(0.5)*600}

Then, the alternative FCT^* computation has been performed as

$$FCT* = \frac{1}{1 + \frac{1}{|POS|} \sum_{t \in POS} fcerror_t^{relNoBias}} \qquad (8.13)$$

Again, this measure sorts out a complete certain part of the demand and controls it by pure MTF. Due to its relatively low practical relevance, a more detailed analysis of this hypothesis is left to further research. So is the determination of finding a way to derive $S\%^*$ and S^* in these environments.

8.11.5 Experimental Setup for the Extension to the Multi-Product Case

Table 8.14

Table 8.14 Comparison of analytical and numerical FCT* derivation for distributions with a median different from zero

Distribution	Analytical derivation		Numerical derivation			
	FCT* via E(/FCerr/)	FCT* via alternative	FCT*	Repl. 1	Repl. 2	Repl. 3
$-0.3 + bernoulli\left(\frac{1}{3}\right) \cdot 0.9$	0.71	0.63	0.60	0.6	0.6	0.6
$-0.6 + bernoulli\left(\frac{2}{3}\right) \cdot 0.9$	0.71	0.63	0.56	0.6	0.6	0.5

8.11.6 Analysis of Drivers for the Impact Magnitude

Factorial Fit: Avg_effect versus TBO; Batchsize; ...

```
Estimated Effects and Coefficients for Avg_effect (coded units)

Term                    Effect     Coef   SE Coef       T       P
Constant                         0.3257   0.03963    8.22   0.000
TBO                     0.1976   0.0988   0.03963    2.49   0.018
Batchsize               0.0568   0.0284   0.03963    0.72   0.479
Demand_var             -0.3455  -0.1728   0.03963   -4.36   0.000
Process_var            -0.0051  -0.0026   0.03963   -0.06   0.949
FC_err                  0.5493   0.2746   0.03963    6.93   0.000
TBO*Batchsize          -0.0536  -0.0268   0.03963   -0.68   0.504
TBO*Demand_var         -0.1918  -0.0959   0.03963   -2.42   0.022
TBO*Process_var         0.0420   0.0210   0.03963    0.53   0.600
TBO*FC_err              0.1464   0.0732   0.03963    1.85   0.075
Batchsize*Demand_var    0.0441   0.0220   0.03963    0.56   0.582
Batchsize*Process_var  -0.2214  -0.1107   0.03963   -2.79   0.009
Batchsize*FC_err        0.0514   0.0257   0.03963    0.65   0.522
Demand_var*Process_var -0.0482  -0.0241   0.03963   -0.61   0.547
Demand_var*FC_err      -0.4143  -0.2071   0.03963   -5.23   0.000
Process_var*FC_err     -0.0022  -0.0011   0.03963   -0.03   0.978


S = 0.266383   R-Sq = 76.20%   R-Sq(adj) = 64.30%


Analysis of Variance for Avg_effect (coded units)

Source               DF   Seq SS   Adj SS   Adj MS       F       P
Main Effects          5  3.90941  4.80432  0.96086   13.54   0.000
2-Way Interactions   10  2.90560  2.90560  0.29056    4.09   0.001
Residual Error       30  2.12880  2.12880  0.07096
  Pure Error         30  2.12880  2.12880  0.07096
Total                45  8.94380


Unusual Observations for Avg_effect

Obs  StdOrder  Avg_effect      Fit   SE Fit  Residual  St Resid
 20        20     0.49500  1.32250  0.18836  -0.82750     -4.39R
 27        27     1.17000  0.66000  0.15380   0.51000      2.34R
 36        36     2.15000  1.32250  0.18836   0.82750      4.39R
 43        43     0.19500  0.66000  0.15380  -0.46500     -2.14R

R denotes an observation with a large standardized residual.


Estimated Coefficients for Avg_effect using data in uncoded units

Term                           Coef
Constant                   -1.35378
TBO                     0.000014192
Batchsize                 0.0498461
Demand_var                 0.391944
Process_var            5.85446E-05
FC_err                      1.46258
TBO*Batchsize          -8.94097E-07
TBO*Demand_var         -6.26702E-05
TBO*Process_var         3.49826E-09
TBO*FC_err              0.000159081
Batchsize*Demand_var    0.00191993
Batchsize*Process_var  -2.45949E-06
Batchsize*FC_err          0.0074426
Demand_var*Process_var -1.05074E-05
Demand_var*FC_err          -1.17724
Process_var*FC_err     -1.58514E-06
```

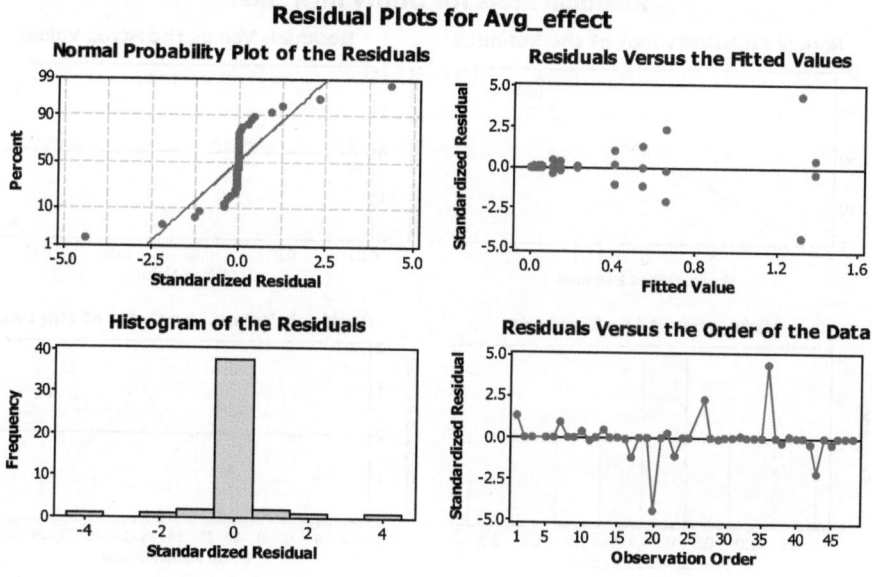

Fig. 8.11 Residual plots for average improvement effect

Figure 8.11

General Linear Model: Utility impr. pull versus D_var; FC_err

```
Factor   Type   Levels  Values
D_var    fixed     2    2; 3
FC_err   fixed     3    1; 2; 3


Analysis of Variance for Utility impr. pull. using Adjusted SS for Tests

Source        DF     Seq SS      Adj SS      Adj MS       F       P
D_var          1   0.0031608   0.0037181   0.0037181   10.69   0.008
FC_err         2   0.0011666   0.0013045   0.0006523    1.88   0.203
D_var*FC_err   2   0.0001394   0.0001394   0.0000697    0.20   0.822
Error         10   0.0034773   0.0034773   0.0003477
Total         15   0.0079440


S = 0.0186477   R-Sq = 56.23%   R-Sq(adj) = 34.34%


Unusual Observations for Utility impr. pull

        Utility
         impr.                                          St
Obs       pull       Fit     SE Fit   Residual   Resid
  1    0.012834  0.012834   0.018648   0.000000     * X

X denotes an observation whose X value gives it large influence.
```

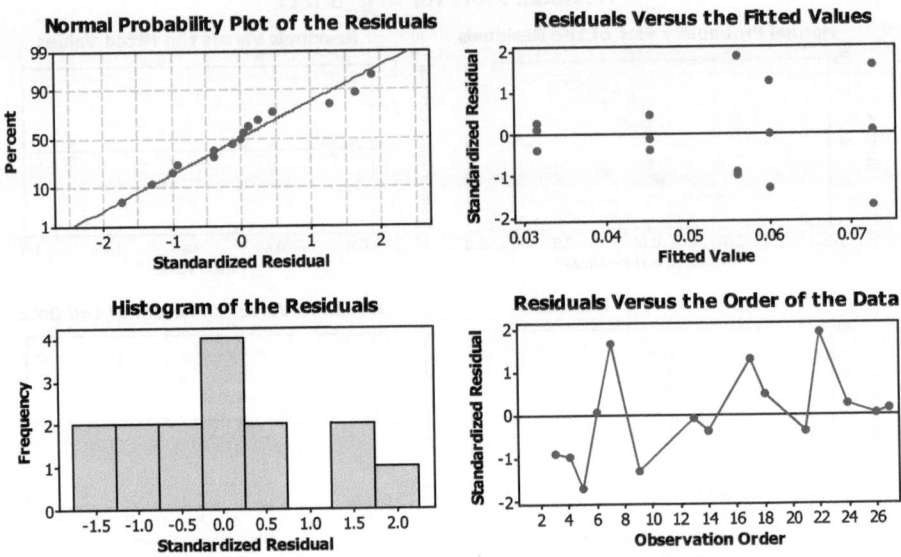

Fig. 8.12 Residual plots for utility improvement compared to pull (Pure MTS)

Figure 8.12

General Linear Model: Utility impr. push versus D_var; FC_err

```
Factor   Type    Levels   Values
D_var    fixed        2   2; 3
FC_err   fixed        3   1; 2; 3


Analysis of Variance for Utility impr. push, using Adjusted SS for Tests

Source       DF   Seq SS    Adj SS   Adj MS       F       P
D_var         1  0.19418   0.05077  0.05077    1.52   0.248
FC_err        2  0.65170   0.68108  0.34054   10.23   0.005
D_var*FC_err  2  0.43506   0.43506  0.21753    6.53   0.018
Error         9  0.29965   0.29965  0.03329
Total        14  1.58058


S = 0.182468   R-Sq = 81.04%   R-Sq(adj) = 70.51%


Unusual Observations for Utility impr. push

        Utility
          impr.
Obs        push       Fit    SE Fit   Residual  St Resid
  1     0.76022   0.76022   0.18247   -0.00000      * X
 22     0.79359   0.49343   0.10535    0.30017     2.01 R


R denotes an observation with a large standardized residual.
X denotes an observation whose X value gives it large influence.
```

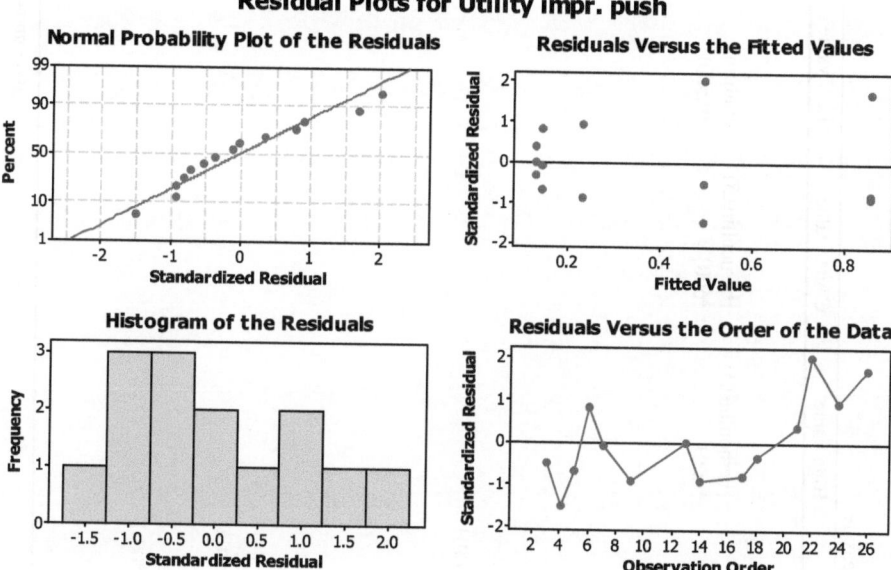

Fig. 8.13 Residual plots for utility improvement compared to push (Pure MTF)

Figure 8.13, Table 8.15

Regression Analysis: Avg Inc versus CDV

```
The regression equation is
Avg Inc = - 0.237 + 0.612 CDV


Predictor        Coef    SE Coef       T        P
Constant     -0.23669    0.04434   -5.34    0.118
CDV           0.61164    0.05817   10.51    0.060


S = 0.0154421   R-Sq = 99.1%   R-Sq(adj) = 98.2%


Analysis of Variance

Source           DF        SS        MS        F        P
Regression        1  0.026362  0.026362   110.55    0.060
Residual Error    1  0.000238  0.000238
Total             2  0.026600
```

8.12 Anylogic Model of Case Study

Figure 8.14

Table 8.15 Parameterization and results of the experiment to derive a demand variability measure

Category	Parameter	Fixed value	High value	Center value	Low value
Demand model	Number of products (m)	1			
	FC error [m]	−0.2+bernoulli(0.5) *0.4			
	Demand distribution [m]		{bernoulli(0.1) *600}	{bernoulli(0.5) *600}	uniform (0,600)
	TBFC	10,000			
	TBO [m]	{5,000}			
	CLT [m]	{0}			
	Order batchsize[m]	{20,20,20}			
Planning segments	Process steps (n)	3			
	Transport batchsizes [n]	{20,20,20}			
	SCT[n,m]	{(1),(3),(1.5)}			
	Transportation [n]	{uniform(20,60), uniform(20,60), uniform(20,60)}			
	C/O time [n]	{0,0,0}			
	Setup family[n,m]	{(0), (0), (0)}			
	Maintenance TTR [n]	{0,0,0}			
	Maintenance TBF [n]	{Inf, Inf, Inf}			
	Scrap PBF [n]	{Inf, Inf, Inf}			
	Rework PBF [n]	{Inf, Inf, Inf}			
	Rework TTR [n]	{0,0,0}			
	Minor Stops PBF [n]	{exponential(0.01), exponential(0.07), exponential (0.01)}			
	Minor Stops TTR [n]	{uniform(6,15), uniform(6,15), uniform(6,15)}			
	Breakdown TBF [n]	{uniform(0,500), uniform(0,500), uniform(0,500)}			
	Breakdown TTR [n]	{150,150,150}			
	Shift Uptime [n]	{960, 1,440, 960}			
	Shift Downtime [n]	{480,0,480}			

(continued)

Table 8.15 (continued)

Category	Parameter	Fixed value	High value	Center value	Low value
Derived factors	Peak demand (in TBFC)		600	600	600
	Average demand		60	300	300
	Demand variability (StdDev)		180	300	173
	Demand CV		3	1	0.58
	Demand Kurtosis		5.10	−2	−1.20
	Demand Skewness		2.67	0	0
Results	Utility hybrid		0.91	0.85	0.90
	Utility pure MTF		0.63	0.65	0.79
	Utility pure MTS		0.76	0.77	0.82
	% Inc. MTF->hybrid		0.46	0.31	0.14
	% Inc. MTS->hybrid		0.21	0.10	0.09
	Avg. increase		0.34	0.21	0.11

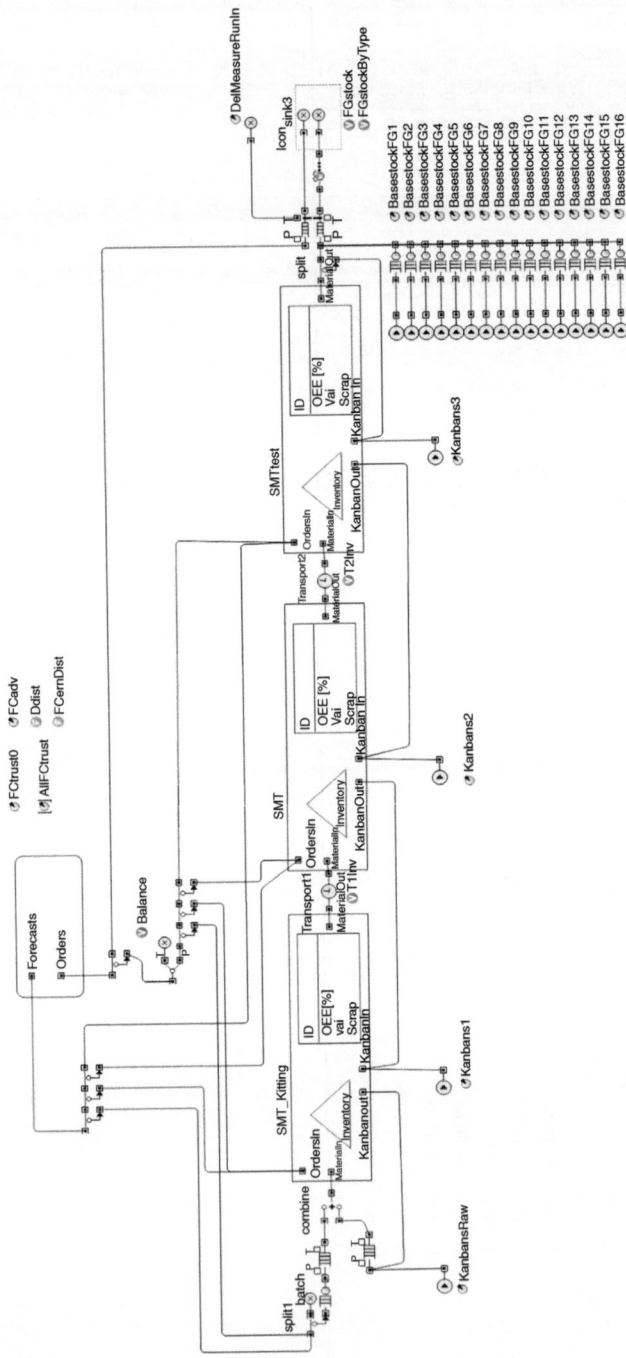

Fig. 8.14 AnyLogic root model of case application

8.13   Used Software Products

AnyLogic, Version 6.4
XJ Technologies Company
Office 410
49 Nepokorennykh pr.
195,220 St. Petersburg
RUSSIAN FEDERATION
http://www.xjtek.com

MiniTab, Version 14
ADDITIVE GmbH
Max-Planck-Str. 22b
61,381 Friedrichsdorf
GERMANY
http://www.minitab.com

Excel 2003
Microsoft Corporation
1 Microsoft Way
Redmond WA
USA
http://www.microsoft.com

8.14 About the Author

Christoph Karrer is passionate about the application of quantitative methods and new technologies to improve the performance of production systems and supply chains. In his professional career, he has successfully designed and implemented production control strategies for five different companies in the High-Tech and Automotive industries.

Christoph holds a PhD in Engineering (Dr.-Ing.) from the Department of Production Management at the Technical University of Berlin and studied Information Engineering and Management (Diplom) at the Karlsruhe Institute of Technology (KIT), Stanford University, and the University of California at Santa Barbara. He currently works as project leader in the Operations Practice of McKinsey & Company in Munich.

References

Abele E, Kluge J, Näher U (2009) Handbuch Globale Produktion. Hanser Fachbuchverlag, München (in German)

Alicke K (2005) Planung und Betrieb von Logistiknetzwerken: Unternehmsübergreifendes Supply Chain Management (VDI-Buch), 2nd edn. Springer, Berlin (in German)

AnyLogic Manual (2009) AnyLogic version 6.4.0. XJ Technologies, St. Petersburg

Arnold D, Furmans K (2009) Materialfluss in Logistiksystemen, 6th edn. Springer, Berlin (in German)

Babai MZ, Dallery Y, Industriel LG (2009) Dynamic versus static control policies in single stage production-inventory systems. Int J Prod Res 47:415–433

Babai M, Syntetos AA, Dallery Y, Nikolopoulos K (2009) Dynamic Re-order Point Inventory Control with Lead-Time Uncertainty: Analysis and Empirical Investigation International Journal of Production Research 47: 2461–2483

Baker KR, Scudder GD (1990) Sequencing with earliness and tardiness penalties: a review. Oper Res 28:22–36

Bamberg G, Coenenberg AG, Krapp M (2008) Betriebswirtschaftliche Entscheidungslehre, 14th edn. Vahlen, München (in German)

Baynat B, Buzacott JA, Dallery Y (2002) Multiproduct kanban-like control systems. Int J Prod Res 40:4225–4255

Beamon BM, Bermudo JM (2000) A hybrid push/pull control algorithm for multistage, multiline production systems. Prod Plan Contr 11:349–356

Bechte W (1988) Theory and practice of load-oriented manufacturing control. Int J Prod Res 26:375–395

Bechte W (1984) Steuerung der Durchlaufzeit durch belastungsorientierte Auftragsfreigabe bei Werkstattfertigung. VDI-Verlag, Düsseldorf (in German)

Benjaafar S, Cooper WL, Mardan S (2010) Production-inventory systems with imperfect advance demand information and updating. Naval Research Logistics 58:88–106

Benton WC, Shin H (1998) Manufacturing planning and control: the evolution of MRP and JIT integration. Eur J Oper Res 110:411–440

Berger JO (1993) Statistical decision theory and Bayesian analysis, 2nd edn. Springer, Berlin

Berkley BJ (1992) A review of the Kanban production control research literature. Prod Oper Manag 1:393–412

Bertrand JWM, Wortmann JC (1981) Production control and information systems for component manufacturing shops. Elsevier Science, New York

Blazewicz J, Ecker KH, Pesch E, Schmidt G, Weglarz J (2001) Scheduling computer and manufacturing processes, 2nd edn. Springer, New York

Bonney MC, Zhang Z, Head MA, Tien CC, Barson RJ (1999) Are push and pull systems really so different? Int J Prod Econ 59:53–64

C. Karrer, *Engineering Production Control Strategies*, Management for Professionals,   171
DOI 10.1007/978-3-642-24142-0, © Springer-Verlag Berlin Heidelberg 2012

Bonvik AM, Gershwin SB (1996) Beyond Kanban: creating and analyzing lean shop floor control policies. In: Manufacturing and Service Operations Management Conference Proceedings, Dartmouth College, The Amos Tuck School, Hannover, New Hampshire, 46–51

Bosch K (1998) Statistik-Taschenbuch, 3rd edn. Oldenbourg, München (in German)

Brucker P (2007) Scheduling algorithms, 5th edn. Springer, Berlin

Buzacott JA (1989) Queuing models of Kanban and MRP controlled production systems. Eng Costs Prod Econ 17:3–20

Buzacott JA, Shanthikumar JG (1992) A general approach for coordinating production in Multiple-Cell manufacturing systems. Prod Oper Manag 1:34–52

Chaouiya C, Liberopoulos G, Dallery Y (2000) The extended kanban control system for production coordination of assembly manufacturing systems. IIE Trans 32:999–1012

Clark AJ, Scarf H (1960) Optimal policies for a multi-echelon inventory problem. Manag Sci 6:475–490

Claudio D, Krishnamurthy A (2009) Kanban-based pull systems with advance demand information. Int J Prod Res 47:3139–3160

Claudio D, Zhang J, Zhang Y (2007) A hybrid inventory control system approach applied to the food industry. In: Proceedings of the 39th conference on Winter simulation: 40 years! The best is yet to come, Washington D.C.

Cochran JK, Kaylani HA (2008) Optimal design of a hybrid push/pull serial manufacturing system with multiple part types. Int J Prod Res 46:949–965

Dallery Y, Liberopoulos G (2000) Extended kanban control system: combining kanban and base stock. IIE Trans 32:369–386

Delquie P, Luo M (1997) A simple trade-off condition for additive multiattribute utility. J Multi-Criteria Decis Anal 6:248–252

Deshmukh AV, Talavage JJ, Barash MM (1998) Complexity in manufacturing systems. Part 1: analysis of static complexity. IIE Trans 30:645–655

Dobberstein M (1998) Control on the verge of chaos – analysis of dynamic production structures. In: Scherer E (ed) Shop floor control, a systems perspective: from deterministic models towards agile operations management. Springer, Berlin

Drew J, McCallum B, Roggenhofer S (2004) Journey to lean: making operational change stick. Palgrave Macmillan, Hampshire

Flapper SDP, Miltenburg GJ, Wijngaard J (1991) Embedding JIT into MRP. Int J Prod Res 29:329–341

Framinan JM, Gonzalez PL, Ruiz-Usano R (2003) The CONWIP production control system: review and research issues. Prod Plan Contr 14:255–265

Framinan JM, Ruiz-Usano R, Leisten R (2001) Sequencing CONWIP flow-shops: analysis and heuristics. Int J Prod Res 39:2735–2749

Framinan JM, Ruiz-Usano R, Leisten R (2000) Input control and dispatching rules in a dynamic CONWIP flow-shop. Int J Prod Res 38:4589–4598

Gaury EGA (2000) Designing pull production control systems: Customization and robustness. Ph.D. thesis, Tilburg University.

Gaury EGA, Kleijnen JPC, Pierreval H (2001) A methodology to customize pull control systems. J Oper Res Soc 52:789–799

Gaury EGA, Pierreval H, Kleijnen JPC (2000) An evolutionary approach to select a pull system among kanban, conwip and hybrid. J Intell Manuf 11:157–167

Gayon JP, Benjaafar S, Vericourt F (2009) Using imperfect advance demand information in production-inventory systems with multiple customer classes. Manuf Serv Oper Manag 11:128–143

Gelbke S (2008) Einsatz von multiagentensystemen im fertigungsmanagement: grundlagen, entwurf und analyse. Vdm Verlag Dr. Müller, Saarbrücken

Geraghty J (2003) An investigation of pull-type production control mechanisms for lean manufacturing environments in the presence of variability in the demand process. Ph.D. Thesis, University of Limerick, Ireland.

Geraghty J, Heavey C (2005) A review and comparison of hybrid and pull-type production control strategies. OR Spectrum 27:435–457

Geraghty J, Heavey C (2004) A comparison of Hybrid Push/Pull and CONWIP/Pull production inventory control policies. Int J Prod Econ 91:75–90

Glassey CR, Resende MGC (1988) Closed-loop job release control for VLSI circuit manufacturing. IEEE Trans Semicond Manuf 1:36–46

Goldratt EM, Fox RE (1986) The race. North River Press Croton-on-Hudson, New York

Goncalves P (2000) The impact of shortages on push-pull production systems. Techreport, Sloan School of Management, Massachusetts Institute of Technology, Cambridge

Grosfeld-Nir A, Magazine M, Vanberkel A (2000) Push and Pull strategies for controlling multistage production systems. Int J Prod Res 38:2361–2375

Gstettner S, Kuhn H (1996) Analysis of production control systems kanban and CONWIP. Int J Prod Res 34:3253–3273

Günther HO, Tempelmeier H (2012) Produktion und Logistik, 9th edn. Springer, Berlin (in German)

Hall RW (1986) Syncro-MRP: combining kanban and MRP, the Yamaha PYMAC system. In: Driving the productivity machine: production planning and control in Japan. American Production & Inventory Control Society, USA

Hirakawa Y (1996) Performance of a multistage hybrid push/pull production control system. Int J Prod Econ 44:129–135

Hodgson TJ, Wang DW (1991a) Optimal hybrid push/pull strategies for a parallel multistage system: Part 1. Int J Prod Res 29:1279–1287

Hodgson TJ, Wang DW (1991b) Optimal hybrid push/pull strategies for a parallel multistage system: Part 2. Int J Prod Res 29:1453–1460

Hopp WJ, Spearman ML (2008) Factory physics, 3rd edn. McGraw-Hill, Boston

Horst R, Pardalos PM (2002) Handbook of global optimization. Kluwer Academic Pub, Dordrecht

Hoshino K (1996) Criterion for choosing ordering policies – between fixed-size and fixed-interval, pull-type and push-type. Int J Prod Econ 44:91–95

Howard RA (1983) The evolution of decision analysis. In: Howard RA (ed) Decision analysis – A collection of readings. Department of Management Science and Engineering, Stanford University, Palo Alto, CA (USA)

Huang HH (2002) Integrated production model in agile manufacturing systems. Int J Adv Manuf Technol 20:515–525

Huang M, Wang DW, Ip WH (1998) A simulation and comparative study of CONWIP Kanban and MRP production control system in a cold rolling plant. Prod Plan Contr 9:803–812

Institute Iacocca (1991) 21st century manufacturing enterprise strategy. Lehigh University, Bethlehem

Jacobs FR, Bendoly E (2003) Enterprise resource planning: developments and directions for operations management research. Eur J Oper Res 146:233–240

Jensen MT (2001) Robust and flexible scheduling with evolutionary computation. Ph.D. Thesis, University of Aarhus.

Karaesmen F, Buzacott JA, Dallery Y (2002) Integrating advance order information in make-to-stock production systems. IIE Trans 34:649–662

Karaesmen F, Liberopoulos G, Dallery Y (2004) The value of advance demand information in production/inventory systems. Ann Oper Res 126:135–157

Keeney RL, Raiffa H (1993) Decisions with multiple objectives: preferences and value trade-offs: preferences and value Trade-Offs. Cambridge University Press, Cambridge

Khoo LP, Lee SG, Yin XF (2001) Agent-based multiple shop floor manufacturing scheduler. Int J Prod Res 39:3023–3040

Kilsun K, Chhajed D, Paleak US (2002) A comparative study of the performance of push and pull systems in the presence of emergency orders. Int J Prod Res 40:1627–1646

Kimura O, Terada H (1981) Design and analysis of pull system, a method of multi-stage production control. Int J Prod Res 19:241–253

Kleijnen JPC, Gaury EGA (2003) Short-term robustness of production management systems: a case study. Eur J Oper Res 148:452–465

Kleinau P, Thonemann UW (2004) Deriving inventory-control policies with genetic programming. OR Spectr 26:521–546

Koh SG, Bulfin RL (2004) Comparison of DBR with CONWIP in an unbalanced production line with three stations. Int J Prod Res 42:391–404

Kosturiak J, Gregor M (1995) Total production control. Prod Plan Contr 6:490–499

Law AM, Kelton WD (2008) Simulation modeling and analysis, 4th edn. McGraw-Hill, Boston

Lee YH, Lee B (2003) Push-pull production planning of the re-entrant process. Int J Adv Manuf Technol 22:922–931

Lee-Mortimer A (2008) A continuing lean journey: an electronic manufacturer's adopting of Kanban. Assem Autom 28(2):103–112

Li H, Liu L (2006) Production control in a two-stage system. Eur J Oper Res 174:887–904

Li J, Meerkov SM (2009) Production systems engineering. Springer, New York

Li L, Fonseca DJ, Chen DS (2006) Earliness-tardiness production planning for just-in-time manufacturing: a unifying approach by goal programming. Eur J Oper Res 175:508–515

Liberopoulos G (2008) On the trade-off between optimal order-base-stock levels and demand lead-times. Eur J Oper Res 190:136–155

Liberopoulos G, Chronis A, Koukoumialos S (2003) Base stock policies with some unreliable advance demand information. In: Proceedings of the 4th Aegean International Conference on Analysis of Manufacturing Systems, Samos Island.

Liberopoulos G, Koukoumialos S (2008) On the effect of variability and uncertainty in advance demand information on the performance of a make-to-stock supplier. MIBES Transactions International Journal 2 (1): 95–114

Liberopoulos G, Koukoumialos S (2005) Trade-offs between base stock levels, numbers of kanbans, and planned supply lead times in production/inventory systems with advance demand information. Int J Prod Econ 96:213–232

Liberopoulos G, Dallery Y (2000) A unified framework for pull control mechanisms in multi-stage manufacturing systems. Ann Oper Res 93:325–355

Lödding H (2008) Verfahren der Fertigungssteuerung. Grundlagen, Beschreibung, Konfiguration. Springer, Berlin (in German)

Maes J, VanWassenhove LN (1991) Functionalities of production-inventory control systems. Prod Plan Contr 2:219–227

Masin M, Prabhu V (2009) AWIP: a simulation-based feedback control algorithm for scalable design of self-regulating production control systems. IIE Trans 41:120–133

Mitra D, Mitrani I (1989) Control and coordination policies for systems with buffers. Perform Eval Rev 17:156–164

Montgomery DC (2009) Design and analysis of experiments, 7th edn. Wiley, New York

Muchiri P, Pintelon L (2008) Performance measurement using overall equipment effectiveness (OEE): literature review and practical application discussion. Int J Prod Res 46:1–19

Mönch L (2006) Agentenbasierte Produktionssteuerung komplexer Produktionssysteme. Ein agentenbasierter Ansatz. Vieweg+Teubner, Braunschweig (in German)

Mönch L, Stehli M, Zimmermann J, Habenicht I (2006) The FABMAS multi-agent-system prototype for production control of water fabs: design, implementation and performance assessment. Prod Plan Contr 17:701–716

Neumann K, Morlock M (2002) Operations research, 2nd edn. Hanser, München

Nyhuis P, Wiendahl HP (1999) Logistische Kennlinien. Springer, Berlin (in German)

Ohno T (1988) Toyota production system: beyond large-cale production. Productivity Press, New York

Olhager J (2003) Strategic positioning of the order penetration point. Int J Prod Econ 85:319–329

Olhager J, Ostlund B (1990) An integrated push-pull manufacturing strategy. Eur J Oper Res 45:135–142

Orlicky J (1975) Material requirements planning: the new way of life in production and inventory management. McGraw-Hill, New York

Ozbayrak M, Akgun M, Turker AK (2004) Activity-based cost estimation in a push/pull advanced manufacturing system. Int J Prod Econ 87:49–65

Ozbayrak M, Papadopoulou TC, Samaras E (2006) A flexible and adaptable planning and control system for an MTO supply chain. Robotics Comp Intergrated Manuf 22:557–565

Pandy PC, Khokhajaikiat P (1996) Performance modeling of multistage production systems operating under hybrid push/pull control. Int J Prod Econ 43:17–28

Papadopoulou TC, Mousavit A (2007) Dynamic job-shop lean scheduling and conwip shop-floor control using software agents. In: Proceedings. IET International Conference on Agile Manufacturing ICAM 2007, Durham

Perona M, Portioli A (1996) An enhanced loading model for the probabilistic workload control under workload imbalance. Prod Plan Contr 7:68–78

Persentili E, Alptekin E (2000) Product flexibility in selecting manufaturing planning and control. Int J Prod Res 38:2011–2021

Pine BJ (1993) Mass customization: the new frontier in business competition. Harvard Business School Press, Boston

Pinedo M (2008) Scheduling: theory, algorithms, and systems, 3rd edn. Prentice Hall, Upper Saddle River

Prasad RP (1997) Surface mount technology: principles and practice, 2nd edn. Springer, New York

Pyke DF, Cohen MA (1990) Push and pull in manufacturing and distribution systems. J Oper Manag 9:24–43

Razmi J, Rahnejat H, Khan MK (1998) The application of analytic hierarchy process in classification of material planning and control systems. Int J Oper Prod Manag 18:1134–1151

Razmi J, Rahnejat H, Khan MK (1996) A model to define hybrid systems at the interface between push and pull systems. In: Khan MK (ed) Advanced Manufacturing Processes, Systems and Technologies (AMPST 96). Wiley, New York

Rother M, Shook J (1998) Learning to see. Lean Enterprise Institute, Brookline

Rücker T (2006) Optimale Materialflusssteuerung in heterogenen Produktionssystemen. Deutscher Universitätsverlag, Wiesbaden (in German)

Sabuncuoglu I, Bayiz M (2000) Analysis of reactive scheduling problems in a job shop environment. Eur J Oper Res 126:567–586

Sage AP, Armstrong JE (2000) Introduction to systems engineering, 1st edn. Wiley, New York

Sage AP, Rouse WB (1999) Handbook of systems engineering and management. Wiley, New York

Sarker BR, Fitzsimmons JA (1989) The performance of push and pull systems: a simulation and comparative study. Int J Prod Res 27:1715–1732

Scheer AW (2007) CIM Computer Integrated Manufacturing: Der computergesteuerte Industriebetrieb, 4th edn. Springer, Berlin

Scherer E (1998) Models, systems and reality: knowledge generation and strategies for systems design. In: Scherer E (ed) Shop floor control, a systems perspective: from deterministic models towards agile operations management. Springer, Berlin

Schneeweiß C (2002) Einführung in die Produktionswirtschaft, 8th edn. Springer, Berlin (in German)

Scholz-Reiter B, Philipp T, DeBeer C, Windt K, Freitag M (2006) Einfluss der strukturellen Komplexität auf den Einsatz von selbststeuernden logistischen Prozessen. In: Pfohl, HC, Wimmer, T (eds) Steuerung von Logistiksystemen – auf dem Weg zur Selbststeuerung. Konferenzband zum 3. BVL-Wissenschaftssymposium Logistik Deutscher Verkehrs-Verlag, Hamburg (in German).

Schwarzendahl R (1996) The introduction of Kanban principles to material management in EWSD production. Prod Plan Contr 7:212–221

Sharma S, Agrawal N (2009) Selection of a pull production control policy under different demand situations for a manufacturing system by AHP-algorithm. Comput Oper Res 36:1622–1632

Simchi-Levi D, Kaminsky P, Simchi-Levi E (2007) Designing and managing the suppy chain, 3rd edn. McGraw-Hill, Boston

Slomp J, Bokhorst JAC, Germs R (2009) A lean production control system for high-variety/low-volume environments: a case study implementation. Prod Plan Contr 20:586–595

Spearman ML, Woodruff DL, Hopp WJ (1990) CONWIP: a pull alternative to kanban. Int J Prod Res 28:879–894

Spearman ML, Zazanis MA (1992) Push and pull production systems: issues and comparisons. Oper Res 40:521–532

Stadtler H, Kilger C (2008) Supply chain management and advanced planning: concepts, models, software, and case studies, 4th edn. Springer, Berlin

Stocker UM, Waldmann KH (2003) Stochastische Modelle: Eine anwendungsorientierte Einführung. Springer, Berlin (in German)

Stuber F (1998) Approaches to shop floor scheduling – a critical review. In: Scherer E (ed) Shop floor control, a systems perspective: from deterministic models towards agile operations management. Springer, Berlin

Suri R (1998) Quick response manufacturing. Productivity Press, New York

Taguchi G (1986) Introduction to quality engineering: designing quality into products and processes. Quality Resources, New York

Takahashi K, Nakamura N (2002) Decentralized reactive Kanban system. Eur J Oper Res 139:262–276

Takahashi K, Soshiroda M (1996) Comparing integration strategies in production ordering systems. Int J Prod Econ 44:83–89

Tan T, Güllü R, Erkip N (2007) Modeling imperfect advance demand information and analysis of optimal inventory policies. Eur J Oper Res 177:897–923

Tardif V, Maaseidvaag L (2001) An adaptive approach to controlling kanban systems. Eur J Oper Res 132:411–424

Tempelmeier H (2001) Master planning mit Advanced Planning Systems. Books on Demand, Norderstedt (in German)

Treharne JT, Sox CR (2002) Adaptive inventory control for nonstationary demand and partial information. Manag Sci 48:607

Tsubone H, Kuroya T, Matsuura H (1999) The impact of order release strategies on the manufacturing performance for shop floor control. Prod Plan Contr 10:58–66

Vinod V, Sridharan R (2009) Simulation-based metamodels for scheduling a dynamic job shop with sequence-dependent setup times. Int J Prod Res 47:1425–1447

Wang DW, Chen XZ (1996) Experimental push/pull production planning and control system. Prod Plan Cont 7:236–241

Wang DW, Hodgson TJ (1992) Computation reduction for large scale Markov process to analyze multistage production systems. Int J Prod Res 30:2411–2420

Wang DW, Xu CG (1997) Hybrid push/pull production control strategy simulation and its applications. Prod Plan Contr 8:142–151

Weitzman R, Rabinowitz G (2003) Sensitivity of push and pull strategies to information updating rate. Int J Prod Res 41:2057–2074

Womack JP, Jones DT (2003) Lean thinking: banish waste and create wealth in your corporation, revised and updated. Free Press, New York

Womack JP, Jones DT, Roos D (2007) The machine that changed the world: the story of lean production. Free Press, New York

Xiong G, Nyberg TR (2000) Push/pull production plan and schedule used in modern refinery CIMS. Robotics Comp Intergrated Manuf 16:397–410